New Issues in Polar Tourism

Dieter K. Müller • Linda Lundmark
Raynald H. Lemelin
Editors

New Issues in Polar Tourism

Communities, Environments, Politics

Springer

Editors
Dieter K. Müller
Department of Geography and Economic History
Umeå University
Umeå, Sweden

Linda Lundmark
Department of Geography and Economic History
Umeå University
Umeå, Sweden

Raynald H. Lemelin
School of Outdoor Recreation, Parks and Tourism
Lakehead University
Thunder Bay, ON, Canada

ISBN 978-94-007-5883-4 ISBN 978-94-007-5884-1 (eBook)
DOI 10.1007/978-94-007-5884-1
Springer Dordrecht Heidelberg New York London

Library of Congress Control Number: 2012955372

Printed on acid-free paper

Springer is part of Springer Science+Business Media (www.springer.com)

Preface

I felt always attracted by northern environments; thus, it is logical that northern and polar areas also have increasingly and, once again, become of great scientific interests. Located in Umeå, northern Sweden, this is of course also a reasonable thing to do, not least considering the ongoing struggle for developing tourism in order to maintain northern Sweden as an attractive spot to live in. Still, interest is only one factor among many others needed for starting and finalizing a book on Polar Tourism. During my postdoc period at the European Tourism Research Institute (ETOUR) in Östersund, Sweden, I was fortunate to have an employer with considerable experiences in polar issues. Dr. Olle Melander, a physical geographer by training, is an individual with a multitude of experiences in organizing and leading scientific and touristic expeditions to the polar areas. With his experience, he also encouraged and almost requested my professional engagement in the North in particular. He invited me to speak for Swedish polar scientists at the annual Polar Forum of the Swedish Polar Research Secretariat on the need of tourism and social science research. Dr. Melander convinced me about the necessity to further engage into a fascinating field of research. He also persuaded me to organize the 2nd IPTRN conference at the Abisko Research Station, Northern Sweden. This was, as usual, an excellent advice, and thus I would like to use this opportunity to express my gratitude to Olle for all his support, all stories, discussions, and great memories.

I also would like to my coeditors Linda Lundmark sharing my interest for polar tourism during our everyday working life at Umeå University and Harvey Lemelin. Both had to deal with my sudden appointment as dean, which certainly interfered with work on this book. Moreover, life and polar tourism research at Umeå University was certainly improved by Suzanne de la Barre and Patrick Brouder, who I would like to thank for friendship and discussions. Other Umeå colleagues who supported this project in various ways are Roger Marjavaara, Carina Keskitalo, Patrick Lantto, and others at the Department of Geography and Economic History and the strong research environment of Northern Studies at Umeå University. Thanks also to other supporters and friends, Jarkko Saarinen, Rannveig Olofsdottir, Edward Huijbens, C. Michael Hall, Albina Pashkevich, and all those who I forgot to mention, who make polar tourism research great fun. An even greater thank you to my partner Åsa Zetterström, my best friend and travel companion, who shares my interest in northern places.

I would like to acknowledge the financial support from the Swedish Research Council and particularly from MISTRA and the MISTRA Arctic Futures program that enabled me to pursue this book project. Also, Christer Jonasson and all the people at the Abisko Research Station should be thanked for providing great facilities in a great location. Finally, I would like to thank Margaret Deignan for facilitating this great opportunity to publish with Springer.

Finally, we all would also like to thank all contributors of this volume and all friends within the IPTRN for discussions and good companionship in polar places and for creating virtual polar worlds in the internet.

Umeå, Sweden Dieter K. Müller

Contents

List of Figures

List of Tables

Contributors

Megan R. Balks, Ph.D. Department of Earth and Ocean Sciences, Waikato University, Hamilton, New Zealand

Kees Bastmeijer, Ph.D. Tilburg Law School, Department European and International Public Law, Tilburg University, Tilburg, The Netherlands

Patrick Brouder Department of Geography and Economic History, Umeå University, Umeå, Sweden

Jackie Dawson, Ph.D. Department of Geography, University of Ottawa, Ottawa, Canada

Sandra Magdalena Granquist Institute of Freshwater Fisheries, Hvammstangi, Iceland

The Icelandic Seal Centre, Hvammstangi, Iceland

Heike Herata German Federal Environment Agency, Dessau-Roßlau, Germany

Edward H. Huijbens, Ph.D. Icelandic Tourism Research Centre, University of Akureyri, Akureyri, Iceland

Gunnar Thór Jóhannesson, Ph.D. Institute of Life and Environmental Sciences, Department of Geography and Tourism, University of Iceland, Reykjavík, Iceland

Margaret Johnston, Ph.D. School of Outdoor Recreation, Parks and Tourism, Lakehead University, Thunder Bay, ON, Canada

Manuela Krakau, Ph.D. German Federal Environment Agency, Dessau-Roßlau, Germany

Machiel Lamers, Ph.D. Environmental Policy Group, Wageningen University, Wageningen, The Netherlands

Raynald H. Lemelin, Ph.D. School of Outdoor Recreation, Parks and Tourism, Lakehead University, Thunder Bay, ON, Canada

Jerónimo López-Martínez, Ph.D. Department Geologia y Geoquimica, Facultad de Ciencias, Universidad Autonoma de Madrid, Madrid, Spain

Linda Lundmark, Ph.D. Department of Geography and Economic History, Umeå University, Umeå, Sweden

Patrick T. Maher, Ph.D. Outdoor Recreation and Tourism Management, University of Northern British Columbia, Prince George, BC, Canada

Tuukka Mäkitie Department of Geography, University of Helsinki, Helsinki, Finland

Department of Development and Planning, Aalborg University, Aalborg, Denmark

Dieter K. Müller, Ph.D. Department of Geography and Economic History, Umeå University, Umeå, Sweden

Per Åke Nilsson, Ph.L. Hólar University College, Hólar, Sauðárkrókur, Iceland

The Icelandic Seal Centre, Hvammstangi, Iceland

Hammerdal Förlag & Reportage, Hammerdal, Sweden

Tanya A. O'Neill Landcare Research, Manaaki Whenua, Hamilton, New Zealand

Department of Earth and Ocean Sciences, Waikato University, Hamilton, New Zealand

Outi Rantala, Ph.D. Multidimensional Tourism Institute, Faculty of Social Sciences, University of Lapland, Rovaniemi, Finland

Eric J. Shelton Department of Tourism, School of Business, University of Otago, Dunedin, New Zealand

Emma Stewart, Ph.D. Faculty of Environment, Society & Design, Lincoln University, Lincoln, Canterbury, New Zealand

Anna Thompson-Carr, Ph.D. Department of Tourism, School of Business, University of Otago, Dunedin, New Zealand

Seija Tuulentie, Ph.D. Finnish Forest Research Institute, Rovaniemi, Finland

Arvid Viken Department of Sociology, Political Science and Community Planning, University of Tromsø, Tromsø, Norway

Anna-Liisa Ylisirniö, Ph.D. Arctic Centre, University of Lapland, Rovaniemi, Finland

Chapter 1
Introduction: New Issues in Polar Tourism

Dieter K. Müller, Linda Lundmark, and Raynald H. Lemelin

Abstract During the recent decade, academic interest in polar tourism certainly reached a new peak. The reasons for this development are plentiful. An important reason is of course the growing numbers of tourists arriving in polar regions and the associated attempts of commercial and noncommercial stakeholders to attract even more tourists to polar destinations. This and other changes, such as global climate change, imply new challenges and opportunities for polar regions, warranting research on tourism in the area. This chapter introduces some of the themes discussed in the scientific literature: definition of polar regions, environmental dimensions, and business and community perspectives. Moreover, the chapter reviews some of the institutional changes occurring in polar tourism research. Finally, the chapter provides an overview of the contributions to this research volume.

Keywords Climate change • Community development • Economic development • Polar tourism • Polar regions

During the recent decade, academic interest in polar tourism certainly reached a new peak. The reasons for this development are plentiful. An important reason is of course the growing numbers of tourists arriving in polar regions and the associated attempts of commercial and noncommercial stakeholders to lure even more tourists to polar destinations. This implies new challenges and opportunities for polar regions, warranting research on tourism in the area. Moreover, global

D.K. Müller, Ph.D. (✉) • L. Lundmark, Ph.D.
Department of Geography and Economic History, Umeå University, SE-90781 Umeå, Sweden
e-mail: dieter.muller@geography.umu.se; linda.lundmark@geography.umu.se

R.H. Lemelin, Ph.D.
School of Outdoor Recreation, Parks and Tourism, Lakehead University, Thunder Bay, ON P7B 5E1, Canada
e-mail: rhlemeli@lakeheadu.ca

D.K. Müller et al. (eds.), *New Issues in Polar Tourism: Communities, Environments, Politics*, DOI 10.1007/978-94-007-5884-1_1,

Fig. 1.1 Čuonavággi: Lapporten (the Sami portal), the U-shaped mountain formation dominating the scenery at Abisko, 68°21′N 18°49′E (Photo: Dieter K. Müller)

climate change has caused media attention for the polar regions, where major changes are predicted to happen causing significant impacts on environment and societies. Thus, the last chance to see, for example, polar bears and pristine polar environment generates, in some cases, interest in polar tourism (Lemelin et al. 2010b).

Thus, since the publishing of the first seminal book on polar tourism, *Polar Tourism: Tourism in the Arctic and Antarctic Region*, edited by C.M. Hall and M. Johnston in 1995, an increasing number of research monographs and articles manifest the ongoing increase in interest (Bauer 2001; Maher 2007; Maher and Stewart 2007; Snyder and Stonehouse 2007; Hall and Saarinen 2010a; Lück et al. 2010; Stonehouse and Snyder 2010; Maher et al. 2011b). Among the recent books, *Polar Tourism – A Tool for Regional Development* edited by Grenier and Müller (2011) can be found. This book is based on the first conference of the then newly established International Polar Tourism Research Network (IPTRN) held in 2008 in Kangiqsujuaq, Nunavik. A second conference was held at the Abisko Scientific Research Station, Abisko, Sweden, in June 2010, and this volume documents the outcomes of that meeting (Fig. 1.1).

The IPTRN defines itself as "a group with a shared interest in research that advances the understanding of tourism in and about the Polar Regions. The IPTRN strives to generate, share and disseminate knowledge, resources and perspectives on polar tourism; and strongly supports the development of international collaboration and cooperative relationships between members" (http://www.polartourismnetwork.uqam.ca). Accordingly, this book aims at following this ambition by presenting new and further developing issues in polar tourism. This is achieved not only by theoretical contributions but also by new empirical data from hitherto seldom covered sources and destinations. Thus, this book does not aim to become the final text on polar tourism, but to contribute important pieces to a complex puzzle depicting polar tourism knowledge.

At the conference site, several important dimensions of polar areas can be found, and thus, the location itself represents the situation in many polar settlements in a nutshell. Namely, these are nature conservation, tourism, indigenous population, resource extracting activities, science, and international cooperation. Abisko National Park was among the first national parks to be founded in 1909, preserving a Nordic fell nature and in particular a scenic canyon formed by the Abisko creek. Moreover, the Swedish Tourist Association's (STF) mountain station operated since 1903 (Wall Reinius 2009). It was located at the railway line connecting the iron ore mine in Kiruna with the harbor town Narvik. It also became the final station of Kungsleden (The King's Trail), a 400-km long hiking trail through the Swedish Caledonian mountain range. The tiny village established for the railway construction and maintenance, and today with about 100 inhabitants, has Sami population and is home of active reindeer herders. Finally, Abisko has a long tradition of scientific monitoring of climate change in Arctic areas. Collection of climate data started already in 1913 and today the research station, which is nowadays run by the Swedish Polar Research Secretariat, is part of INTERACT, the International Network for Terrestrial Research and Monitoring in the Arctic. Moreover, research activities in Abisko turned the village into an international destination and a settlement with the highest average educational level in the whole of Sweden.

Thus, the local situation depicts several of the major challenges and opportunities uniting polar places and also shows the complexity signifying polar issues, contrasting often stereotype perceptions of particularly the Arctic as homogenous, backward, and nature only. In the remainder of this introduction these current issues are further presented and justified.

1.1 Entering the Agenda

About 100 years after the race for the poles, polar areas have once again gained significant public attention. This is also manifested by increasing numbers of tourists arriving there (Hall and Saarinen 2010b). Accordingly, an estimated 12 million tourists (excluding tourism to the Russian Arctic) visited the Arctic in 2006. For the Antarctic, roughly 40,000 visitors were estimated in 2009/10.

The touristic interest in the polar regions has certainly multiple reasons. Partly global tourism has increased, and thus, even remote locations outside the main itineraries have been integrated not least in the search for new business opportunities. In this context, the polar areas still play a marginal role considering a total number of 980 million tourist arrivals globally (UNWTO 2012), and thus, the polar regions have so far not caught the interest of, for example, the UNWTO. Still, the rapid growth in the area implied that tourism is increasingly seen as an important industry, contributing to shaping the future of these areas.

Tourism interests are not only seen from outside the areas. A globalizing economy has also created preconditions forcing peripheral areas to search for new industries where employment can be created, and tourism has been frequently identified for being such an industry (Hall 2007; Saarinen 2007; Grenier and Müller 2011). Places like Santa Park in Rovaniemi and Icehotel in Jukkasjärvi have been promoted as successful examples of how tourism can be used to create alternative livelihoods (Grenier 2007; Müller 2011). In this context, tourism is seen as remedy for structural problems of peripheral areas like depopulation and deindustrialization and not least as an opportunity for indigenous populations (Kapashesit et al. 2011). This is certainly applicable for most Arctic destinations, but Hall (2000) even shows the significance of Antarctic tourism on gateway cities in the southern hemisphere.

Besides these economic issues however major interest has been caused by ongoing environmental change in the area (Johnston 2006). Accordingly, climate change in particular has created a growing awareness of changes that are expected to be most significant in polar areas. Thus, issues related to protection of wildlife, adaptation of communities, and codes of conducts have been monitored and discussed in multiple publications (Hall and Saarinen 2010a; Stonehouse and Snyder 2010; Maher et al. 2011b). Particularly, last chance tourism, a consequence of the growing awareness of climate change, has been seen as important reason for visiting the areas (Lemelin et al. 2010a, b). Even the question to what extent polar tourism is contributing to the eradication of its own attraction base has been discussed. In this context, the question whether polar tourists become ambassadors and environmental role models has been debated (Maher 2011).

A further reason for the recent interest in the polar regions is related to the opening of polar oceans for international seaways. Climate change has caused a shortening of the ice season and thus increased accessibility for tourism, but also interests related to trade and resource exploitation. This entailed new geopolitical struggle over sovereignty in the Arctic (Dodds 2010), but also increasingly in the southern Atlantic Ocean (Benwell and Dodds 2011). Timothy (2010) argues that this struggle comprises tourism, too. Thus, he interprets the establishment of new national parks in the Canadian High Arctic as a tool for manifesting national sovereignty and presence.

Last but not least, climate change also entailed a renewed interest in polar sciences. Particularly the International Polar Year (IPY) 2007/2008 marks a hallmark for scientific endeavor. However, only few projects focusing on tourism were included in the IPY-sponsored research activities, despite significant expectations within the polar tourism research community (Stewart et al. 2005; Maher 2007). An indirect impact of the IPY is however the mobilization of tourism researchers interested in polar areas and indeed the establishment of organizations like the IPTRN, its conferences, and other's conferences.

1.2 Defining Polar Areas and Polar Tourism

There are multiple attempts to define polar areas presented in the various books on the topic. This book has no ambition to repeat definitions presented in recent volumes on the issue (cf. Hall and Saarinen 2010b; Maher et al. 2011b). Most of these definitions are based on scientific features like tree line or permafrost. Social science is absent in these definitions, and also the Arctic Human Development Report (AHDR) mainly departs from previous definitions based on science. Thus, regarding socioeconomic characteristics areas in Northern Sweden and Finland may have more in common with British Columbia and Northern Ontario than with areas considered being part of the Canadian Arctic. However meanwhile the former are the northernmost areas in Sweden and Finland, respectively, the latter are southern provinces within Canada. This indicates the sometimes blurry results of a complex political process entailing delineations of the polar areas.

Hence, the ambition of this book is to raise new perspectives on definitional issues. Thus, focus here is rather on critical assessments of taken-for-granted scientific delineations of the polar areas. Instead of simply departing from these definitions, the processes of defining and of knowledge production are at the core of the scientific interest presented here. This has been done with respect to geopolitics (Keskitalo 2004); the role of tourism in this context has so far however been overlooked.

Even regarding polar tourism, multiple definitions are available in the various publications on the issue. Here however definitions are not dependent on environmental delineations solely. Instead, perceptions of polar influence and form of the polar tourism product, which make defining polar tourism a tricky challenge. Obviously certain activities like watching polar bears seem to be genuine polar tourism activities – still they can be accomplished in zoos around the world, too. Moreover, polar attractions like the Fram Museum (Fig. 1.2) in Oslo do not only display the famous vessel but also the history of Norwegian polar explorations. Are visits to this and other museums a part of polar tourism? From this perspective, visitors certainly do have interests with respect to polar issues. Similar links can also be made regarding the various meetings organized to discuss polar issues. Is the annual meeting of International Association of Antarctic Tour Operators (IAATO) in Turin, Italy, in 2010 a part of polar tourism? Once again, participants are certainly traveling to Turin for the sake of polar areas. Thus, Maher et al. (2011a) argues that these forms of tourism should be included in definitions of polar tourism.

In this context, it is important to remember that tourism has to be seen as system linking tourists' origins to destinations (Hall 2005). In addition, most definitions assume that tourists should leave their home environment. With respect for polar tourism, this implies that travel within the polar areas should not necessarily be seen as polar tourism since environments in origin and destination at least should be similar. Thus, polar tourism usually links tourists from a non-polar origin to a polar destination.

Fig. 1.2 Polar heritage: the Fram Museum in Oslo, Norway (Photo: Dieter K. Müller)

Still, definitional problems as discussed by Leiper (2008) and Smith (1998) also apply to polar tourism. This means that demand-led and supply-led definitions may result in different outcomes since the former refers to tourist motives and the latter to the tourist industry, respectively. Thus, demand-led definitions focus on special-(polar)-interest tourists. VFR tourists to polar destinations are excluded since they travel for motives not related to the polar features of their destination. In contrast, supply-led definitions include VFR tourists since they use the same transport, restaurants, and maybe hotels as special-interests tourists. The same applies for business tourists, FIFO workers, and scientists.

Thus, a pragmatic approach and an including definition have to be advocated. Hence, polar tourism can be defined as tourism to polar areas and for the sake of polar areas. A consistent definition is however hard to maintain since data does not always allow for distinguishing different tourist motives.

1.3 Tourism in Polar Environments

Besides definitional questions, environmental issues continue to dominate the agenda of tourism in polar regions (cf. Hall and Saarinen 2010a, b; Stonehouse and Snyder 2010; Maher et al. 2011a, b). Often studies address the overall impact of global climate change on resources for tourism like wildlife (Dawson et al. 2010), on risks for biological invasions (Hall 2010), and on tourism business (Brouder and Lundmark 2011; Tervo-Kankare and Saarinen 2011).

Often however focus is more on local emissions. Obviously, a growing interest for polar tourism amplifies the potential impact on the environment. This is particularly the case since tourism is concentrated to a few spots mainly.

In the Antarctic in particular, the Antarctic Peninsula has become a hotspot of tourism development (Bauer 2001). Landings take not the least place where cultural heritage related to early explorations is available, too (Roura 2010). Thus, only small spots are exposed to significant disturbance; meanwhile, for large parts of polar areas, direct impact is limited. Nevertheless, in the fragile environment, even small impact tends to be long-lasting (O'Neill et al., Chap. 6, this volume). Similarly in the Arctic, ship-based tourism mainly includes landings in small communities not equipped to host large amount of visitors. Moreover, iconic attractions like polar bears attract visitors increasingly and thus require thorough management (Lemelin et al. 2010a).

In many cases however major risks are induced by the increased presence of human activities in a fragile environment. For example, Stonehouse and Snyder (2010) list the following environmental challenges for sea-based tourism in polar environments: oil spills, sewage and wastewater, solid waste, and threats to wildlife. However, not all studies confirm these risks clearly since wildlife sometimes does not seem to be disturbed by tourism. Still precautionary action is usually forwarded (Malcolm and Penner 2011; van Polanen Petel 2011).

The concern for the polar environment and the impacts caused by global and local processes also call for management and governance (Lamers 2009). In the Antarctic, this is closely related to the Antarctic Treaty system (Baastmeijer 2011) but also manifested in guidelines adopted by tour operators (Stonehouse and Snyder 2010). In contrast, the Arctic is more contested, owing to various geopolitical and economic interests. International regulations are thus weaker and comprise mainly international conventions and agreements (Stonehouse and Snyder 2010). Instead national legislation governs human-environment relations, creating various outcomes in the Northern States.

1.4 Polar Tourism Business and Community Development

The recent debate on tourism and sustainable development shifted focus from not only comprising the environment but also economic and sociocultural dimensions. And indeed tourism development has frequently been identified as government response to declining populations, out-migration, and dependence on outside support (Jenkins et al. 1998; Grenier and Müller 2011).

Although successful examples are in place, many peripheral places in the North still suffer from bad accessibility and other factors related to their locations.

Certainly the sometimes isolated characteristics of polar communities imply a specific attractiveness, but this comes with a price. High costs for tourism products sometimes in combination with doubtful quality not least regarding food and accommodation services certainly limit the degree to that the tourism potential so far could be utilized.

Moreover, the small scale of many tourism operations implies high costs to actually become visible on a global market. Thus, it is the mass tourism destinations that actually contribute significantly to regional development (Lundmark 2006; Müller 2011). Thus, it is questionable whether tourism development is a viable strategy for all polar communities (Müller and Jansson 2007). Many small places do however continue to engage in activities and try to attract visitors.

Thus, challenges to local communities often originate from mass tourism in polar areas. In this context, cruise ships operating in the polar areas offer a travel package that can easily be purchased. This means however that impacts are highly concentrated in space and time. Receiving a cruise ship therefore sometimes overstrains the capacity of small communities to actually host the visitors and draw advantages from their presence.

It is obvious that communities struggle with a multitude of issues related to tourism development. These are sometimes related to management issues on the local level. More often, however, challenges are caused by the fact that polar places increasingly are tied into global processes of political economies. Hence, this book offers a number of perspectives on current challenges for Northern Communities. These are related to attempts of coping with, adapting to, and resisting against change caused by changes in the South. An important question is to what extent polar communities remain targets of global change or whether they manage to gain agency in order to influence future development.

1.5 Outline of This Book

The remainder of this book is structured in three parts further developing the topics raised in this chapter.

1.5.1 Conceptualizing Polar Tourism and Polar Regions

Hall and Saarinen (2010a:449) note that "Tourism always occurs somewhere. Therefore, identifying regional boundaries becomes essential to the categorization of tourists." Defining borders automatically contributes to assigning meaning to the defined area. Thus, different definitions of the polar regions imply different meanings. Particularly during recent years, the conceptualization of polar tourism has entailed great scientific attention (e.g., Hall and Saarinen 2010b; Stonehouse and

Snyder 2010; Maher 2011; Grenier and Müller 2011). Still the variety of forms and expressions, as well as preconditions and consequences, certainly justify a further examination as it is done in this volume.

For the Antarctic boundaries are seldom contested. The Antarctic Treaty 1959 defines the area south of the 60°S latitude as Antarctic. The sub-Antarctic is the zone between 60°S and 45°S latitude. This implies that besides the continent South Ocean islands are included, too. However, biogeographical delineations are sometimes used to define the southern polar region (Hall and Saarinen 2010b). Since the area is largely uninhabited and strongly regulated by, for example, the Antarctic Treaty, dispute about definition is rare.

In contrast, the situation in the northern polar region is different. Here the Arctic Ocean forms the core of the region, which is framed by inhabited land. Hall and Saarinen (2010b) provide a survey of different definitions of the Arctic. Accordingly, most of the definitions are based on natural features of the area. Thus, the Arctic is seen as a homogenous region disrespecting the presence of people in the area; tourism to the Arctic is not tourism to an inhabited place in the first hand. Instead the social construction of the destination is primarily built on biophysical features. Hence, it is just a place different and remote from the surrounding south. Moreover, it is homogenized disregarding contrasting physical settings and living conditions within the area.

Only the definition given by the AHDR is not using isotherms representing physical conditions directly for delineating the Arctic. Still the AHDR (2004:17) states that "There is nothing intuitively obvious about the idea of treating the Arctic as a distinct region." Because of pragmatic reasons, political entities are used roughly, mirroring earlier areas of scientific assessment, namely, the Arctic Monitoring and Assessment Programmes (AMAP), once again falling back on science definitions.

Although aware of the problems regarding the definition of the Arctic as a region, the AHDR attempts to do so anyway. In texts, illustrations, and maps, this is done without acknowledging major linkages to areas outside the defined region. Thus, the Arctic is seen in the tradition of regional geography and Hettner's (1927) Länderkunde highlighting the region as a major focus of scientific inquiry and concerned with the relationship of the physical environment and the human adaptation to it and transformation of it. Environmental determinism seems thus to be inherent of many definitions of the Arctic.

The delineation of the sub-Arctic creates even greater problems, but is on the other hand hardly addressed at all. Usually included in the term polar regions, a major discussion on its specific characteristics is often but not always absent (Hall and Johnston 1995; Maher et al. 2011a, b).

Hence, definitions of the polar regions provide little guidance for discussing polar tourism besides the fact that it is tourism to a different bioregion. From a touristic point of view this means that current definitions of the polar regions mainly correspond to ideas of exotism, depicting the areas as something else. For example, Jacobson (1997) delivers the story of the North Cape and how it is constructed as a tourism attraction. With respect to present populations, this relates to discourses of "othering" northern peoples (cf. Eriksson 2010). Moreover, it is in line with cultural histories of explorations and literary Arctic discourses (Bravo and Sörlin 2002; Jølle and Drivenes 2006; Hansson and Norberg 2009; Ryall et al. 2010) and more recent geopolitical struggle (Keskitalo 2004). In this context, notions of asymmetrical power relations as expressed by Foucault or Said are of course present.

The process and its result can be critiqued as done by Viken (Chap. 3), but still the Arctic is there because of practices and economic realities that position the area in relation to southern centers of population and power. And indeed, owing not least to the scientific concern for climate change and its consequences for geopolitics, the Arctic certainly is a well-embraced construction among politicians, but also tourists, as, for example, the sacralization of the Arctic Circle impressively manifests (Löytynoja 2008).

In the following chapter (Chap. 2), Maher presents a review of polar tourism research. He departs from defining the area, but soon widens the concept of polar tourism toward including travel even outside the defined area. Thus, polar tourism is not only traveling to the polar regions but also tourism for the sake of the polar regions. The latter includes tourism to polar museums or events focusing on the polar regions. Maher then moves on to provide an account of past and current research activities on polar tourism. In this context, he also discusses the relation of research to industry and community development. Accordingly, the lack of attention given to tourism during the recent IPY was not because of a rudimentary knowledge base, but rather owing to other considerations. Still the IPY did provide some platforms (albeit limited) to address tourism in some jurisdictions (i.e., Europe), while virtually ignoring the field of tourism in other jurisdiction (i.e., Canada). Still the IPY contributed anyway to the development of tourism research. This was because of the attention that it caused and the platforms that suddenly were provided to meet with researchers and industry. As a result of this, preconditions for progressing polar tourism research are better now than ever before.

Even Viken's Chap. 3 addresses the polar tourism research and its development. However, Viken takes a critical approach questioning the hegemonies of scientific production. Accordingly most research is published in journals and book series in the south and particularly in the UK and the USA. The consequence is a dominance of outside perspectives on polar regions implying inaccuracies, generalizations, and stereotypes. At the same time, local voices tend to be marginalized and forced to adapt to standards developed within the hegemonic discourses. Viken's solution to the problem is reflexivity and the development of alternative publishing channels, breaking the hegemony of Anglo-American researchers and publishers.

The final chapter (Chap. 4) of this part puts a different twist on definitions of polar. Sheldon departs from Hollinshead's notion of the worldmaking potential of tourism in his discussion of sub-Antarctic islands. He notes that those are often seen just as an extension of the Antarctic and reports of initiatives to establish the sub-Antarctic as a circumpolar region distinct from the Antarctic. To Shelton, this becomes a matter of narrative identities and imaginaries. On the cruise ships toward the sub-Antarctic islands, narratives already provide a plot for place-making and tourist performance.

1.5.2 Politics and the Environment

This section on politics and the environment begins with an examination of how worldviews affect policy and governance and recommends that philosophical underpinnings should be recognized and acknowledge, especially in the context of Antarctica and even the Arctic where multilateral agreements and cooperation are required (Chap. 5). Following this chapter are two case studies both conducted in Antarctica, where management approaches and policies have been effectively implemented to reduce and minimize anthropogenic impacts in the circumpolar south. The last chapter describes how an international tourism cooperative network was established to monitor and develop proactive management strategies in various wildlife tourism destinations in the Arctic.

The chapter by Bastmeijer and Lamers (Chap. 5) discusses how different interpretations of nature by the Consultative Parties have resulted in different interpretations into what types of tourism activities have been deemed acceptable and unacceptable in Antarctica. According to the authors, it is most likely that Consultative Parties "with different human-nature relationships have different views on what the specific "values" of Antarctica are and how these values could best be protected, and, consequently, it is also most likely that these CPs have different opinions on what norms should be set in respect of specific tourism developments." The authors go on to state that attitudes toward nature, and in this case Antarctica, influences governance and management strategies because these attitudes affect how conservation and protection strategies are justified and, in some cases, implemented. A closer look at tourism activities (existing and proposed) reveals that underneath the relatively neutral positions of some Consultative Parties, "substantially different human-nature relationship attitudes may be hidden." These worldviews have influenced and will continue to influence, as the next two chapters describe, how tourism is perceived and managed in Antarctica.

In their overview of the effectiveness of Environmental Impact Assessments in decommissioned (Cape Roberts ice-free storage area, Greenpeace World Park Base, The New Zealand Antarctic Program K123 Loop Moraine campsite, Vanda Station) and active sites (Taylor Valley Visitor Zone) in the Ross Sea Region of Antarctica, O'Neill, Balks, and López-Martinez suggest that decommissioning and remediation strategies have effectively reduced and in some cases virtually eliminated anthropogenic activities in these sites (Chap. 6). Of particular importance is the study conducted in the Taylor Valley Visitor Zone, which suggests that impacts from tourism activities, with proper management, can be minimized and localized. These conclusions are based on the visitor site assessment schemes (VISTA), which aims to support the Environmental Impact Assessments process and address the shortfall of information on the cumulative impacts of visitor activity through a site monitoring program.

Pursuant to the Act Implementing the Protocol on Environmental Protection to the Antarctic Treaty (AIEP 1998), any activity whether related to scientific research, tourism, or journalism that is organized in Germany or begins on German territory is subject to authorization by the German Federal Environment Agency (Umweltbundesamt, UBA). This reporting mechanism, suggests Krakau and Herata, are to prevent overuse and destruction of key sites in Antarctica (Chap. 7). Although the approach was not designed to manage impacts from tourism, it has nevertheless been relatively effective in minimizing negative impacts for certain tourism activities, including restricting visitor access to the penguin colony at Hanna Point. How this reporting mechanism will be able to address new challenges like personal yachts accessing the area and new types of tourism activities in Antarctica, suggest Krakau and Herata, is still open to debate. Considering that these activities are already occurring, the authors advise that the time to discuss and incorporate these activities within existing policies and management strategies is now.

The last chapter in this section examines how network cooperation has been implemented. By using the Triple Helix Model, the Wild North project has fostered collaboration between academia, governments, and industry in various locals located throughout the Arctic (Chap. 8). Through this approach, wildlife tourism initiatives in Greenland, Iceland, the Faroe Islands, and Norway are now monitoring the impacts from wildlife tourism; they have developed a code of conduct, and they will soon be incorporating local communities in the process.

1.5.3 Business and Community Perspectives

Tourism in the Arctic has for long been established as one important avenue for regional development by governments, and adding to this, an increasing number of tourists also travel to these areas. Tourism development issues and the impact of tourism in sparsely populated and often peripheral and marginal areas are therefore topics in need for further research, not only as case studies but also regarding the theoretical contribution that these areas offer. With the content in this section, business and community perspectives under these conditions in the Arctic are addressed, each chapter contributing to relevant theoretical perspectives. The six chapters in this section address in various ways aspects affecting tourism development, such as changing global economic and mobility patterns, and climate change and social and cultural change occurring with time, as well as the struggle of local people and communities for sustaining and developing livelihoods in remote and unique circumstances in relation to a global tourism market.

The first two chapters of this section are addressing the factors of economic crises and climate change for tourism development in terms of policy and strategies as well as in institutions for this development. In the contribution by Jóhannesson and Huijbens, *Tourism Resolving Crisis? Exploring Tourism Development in Iceland in the Wake of Economic Recession* (Chap. 9), the role of tourism for a nation in economic crisis is in focus. The economic situation in many countries at the end of the last decade was deteriorating and the consequences were far more profound and far-reaching than anyone could have expected. When Iceland was hit by the financial crash in 2008, great hopes for the tourism industry to reduce the impact to the nation's economy were articulated. Jóhannesson and Huijbens reveal that the policy and strategies of Iceland again shifted away from a focus on natural resource extraction, intensifying the focus on the production of Iceland as a tourist destination. This shift, it is argued, is a recurring historical pattern in the Icelandic policy.

In Chap. 10 *A (Ski) Trip Into the Future: Climate Change and Winter Tourism in Polar Sweden in 2040* by Brouder and Lundmark, the changing conditions for tourism entrepreneurship introduced by climate change is in focus. Here, tourism entrepreneurship and climate change is discussed with special reference to winter tourism in the polar region of Sweden, and different scenarios are sketched. An Arctic Tourism Innovation System (ATIS) is discussed stressing the special characteristics of the Arctic for tourism development and adaptation to climate change in particular, highlighting the complementary relationship between entrepreneurs and the institutions that regulate their business.

Chapters 11, 12, and 13 are focusing on how culture and identity can be important for how the development of resources is perceived and discussed by local inhabitants. In Chap. 11, *Conflicting Discourses in Tourism Development: A Case of Kilpisjärvi Village*, the focus is on the physical development of nature-based tourism in a small village of 100 permanent inhabitants in Northern Finland. A recent change of tourists from traditional cross country skiers from Finland to younger Norwegian snowmobilers with a parallel development in the built facilities to upscale second homes has also taken place, causing controversy in the local community. From the interviews, six discourses were distinguished among the residents and others interviewed. The conflicting discourses of economic development and sustaining nature and scenery as pristine as possible were highlighted as the most prominent, as well as reindeer herding. Some interviewees also claimed that the change in economic focus toward tourism had influenced the way they perceived their home village. One important conclusion is that tourism development should be regarded as a social web of discourses, thus supporting a community perspective and the importance of multiple-voice participation.

Will "Free Entry into the Forest" Remain? is the topic of Chap. 12. In this chapter, "the right to roam," the "Right of Public Access," or "everyman's right" is highlighted. The situations and conflicts caused by commercial activities, like berry-picking and nature-based tourism, where entrepreneurs are using the practice of everyman's right on someone else's land, has increased recently. Using blogs from Swedish and Finnish websites, the theme of "us" and "them" comes up as important for the conflicts, where the "othering" of, for example, Asian berry pickers and culturally different tourists are questioned because of their perceived lack of knowledge on how to practice "everyman's right." One important question put forward in this chapter is what will happen to "everyman's right" now when a very mobile modern population and new possibilities for non-extractive economic activity causes conflicts?

The contribution by Müller (Chap. 13), *National Parks for Tourism Development in Subpolar Areas: Curse or Blessing?*, describes a situation where local resistance to a national park establishment is played out. Interestingly, while tourism often is claimed to be one option for the periphery to create development and growth, it is often perceived differently by the local community. In this Swedish case, Müller sets out to analyze the discourses concerning the establishment of a national park in the Vindelfjällen mountain area. Using a qualitative and thematic analysis of selected newspapers, the author finds that some themes are recurring. For the indigenous population, a national park designation could mean that mining which is considered as negative for reindeer herding among other things would be prohibited. On the other hand, many cited a national park designation as a dead hand, hindering not only development options in the area but also local practices such as hunting. Another important theme is mistrust of the authorities from the local community. This came from a core – periphery relations.

The last chapter (Chap. 14) in this section is written by Lemelin et al., *Indigenous People: Discussing the Forgotten Dimension of Dark Tourism and Battlefield Tourism*. This chapter is turning the focus to the indigenous dimensions of dark tourism and battlefield tourism with a case from Canada and a case from the USA. The authors argue that indigenous peoples often are disregarded and misunderstood both in research and in the development and rhetoric of battlefield tourism. The contested meanings of sites of warfare, remembrance, or historic events are rarely explored in literature, contributing to what the authors refer to as a cultural dissonance. This means the exclusion of important themes in research as well as a practical exclusion of marginalized populations from having these sites. The authors argue that it is important to acknowledge indigenous sites of warfare, remembrance, or historic events because it could contribute to a positive process of transformation, healing, and empowerment wherein the past could be turned into history and the voices of indigenous peoples could come forth.

Finally, a conclusion draws together the findings of this book. Though considerable progress has been made, research on polar tourism still is in its infancy. However, rapid global change implies further need for knowledge about the impacts of tourism in polar areas, but also about the motivations attracting tourists to travel to Arctic and Antarctic areas. This book is a step into this direction.

References

AHDR (Arctic Human Development Report) (2004) Stefansson Arctic Institute, Akureyri

AIEP (1998) The protocol on environmental protection to the Antarctic treaty

Baastmeijer K (2011) A long-term strategy for Antarctic tourism: the key to decision making within the Antarctic Treaty System. In: Maher PT, Stewart EJ, Lück M (eds) Polar tourism: human, environmental and governance dimensions. Cognizant Communication Corporation, New York, pp 167–183

Bauer TG (2001) Tourism in the Antarctic: opportunities, constraints and future prospects. Haworth Hospitality, New York

Benwell MC, Dodds K (2011) Argentine territorial nationalism revisited: the Malvinas/Falklands dispute and geographies of everyday nationalism. Pol Geog 30(8):441–449

Bravo MT, Sörlin S (eds) (2002) Narrating the Arctic: a cultural history of Nordic scientific practices. Science History Publications, Nantucket

Brouder P, Lundmark L (2011) Climate change in Northern Sweden: intra-regional perceptions of vulnerability among winter-oriented tourism businesses. J Sustain Tour 19(8):919–933

Dawson J, Stewart E, Scott D (2010) Climate change and polar bear viewing: a case study of visitor demand, carbon emissions and mitigation in Churchill, Canada. In: Hall CM, Saarinen J (eds) Tourism and change in polar regions: climate, environment and experience. Routledge, London, pp 89–103

Dodds K (2010) Flag planting and finger pointing: the law of the Sea, the arctic and the political geographies of the outer continental shelf. Political Geogr 29(2):63–73

Eriksson M (2010) (Re)producing a periphery: popular representations of the Swedish North. Department of Social and Economic Geography, Umeå

Grenier AA (2007) The diversity of polar tourism: some challenges facing the industry in Rovaniemi, Finland. Polar Geogr 30(1–2):55–72

Grenier AA, Müller DK (eds) (2011) Polar tourism: a tool for regional development. Presse de l'Universite de Quebec, Montreal

Hall CM (2000) The tourist and economic significance of Antarctic travel in Australian and New Zealand Antarctic gateway cities. Tour Hosp Res 2(2):157–169

Hall CM (2005) Tourism: rethinking the social science of mobility. Prentice Hall, Harlow

Hall CM (2007) North-south perspectives on tourism, regional development and peripheral area. In: Müller DK, Jansson B (eds) Tourism in peripheries: perspectives from the far north and south. CABI, Wallingford, pp 19–37

Hall CM (2010) Tourism and environmental change in polar regions: impacts, climate change and biological invasion. In: Hall CM, Saarinen J (eds) Tourism and change in polar regions: climate, environment and experience. Routledge, London, pp 42–70

Hall CM, Johnston ME (eds) (1995) Polar tourism: tourism in the Arctic and Antarctic region. Wiley, Chichester

Hall CM, Saarinen J (eds) (2010a) Tourism and change in polar regions: climate, environment and experience. Routledge, London

Hall CM, Saarinen J (2010b) Tourism and change in polar regions: introduction – definitions, locations, places and dimensions. In: Hall CM, Saarinen J (eds) Tourism and change in polar regions: climate, environment and experience. Routledge, London, pp 1–41

Hansson H, Norberg C (eds) (2009) Cold matters: cultural perceptions of snow, ice and cold. Royal Skyttean Society, Umeå

Hettner A (1927) Die Geographie, ihre Geschichte, ihr Wesen und ihre Methoden. Ferdinand Hirt, Breslau

Jacobsen JKS (1997) The making of an attraction: the case of North Cape. Ann Tourism Res 24(2):341–356

Jenkins JM, Hall CM, Troughton M (1998) The restructuring of rural economies: rural tourism and recreation as a government response. In: Butler R, Hall CM, Jenkins J (eds) Tourism and recreation in rural areas. Wiley, Chichester, pp 43–67

Johnston ME (2006) Impacts of global environmental change on tourism in the polar regions. In: Gössling S, Hall CM (eds) Tourism & global environmental change: ecological, social, economic and political interrelationships. Routledge, London, pp 37–53

Jølle HD, Drivenes E-A (eds) (2006) Into the ice: the History of Norway and the Polar Regions. Gyldendal akademisk, Oslo

Kapashesit R, Lemelin RH, Bennett N, Williams G (2011) The Cree ecolodge: success through community empowerment. In: Grenier AA, Müller DK (eds) Polar tourism: a tool for regional development. Presse de l'Universite de Quebec, Montreal, pp 155–178

Keskitalo ECH (2004) Negotiating the Arctic: the construction of an international region. Routledge, London

Lamers M (2009) The future of tourism in Antarctica: challenges for sustainability. Universitaire Pers Maastricht, Maastricht

Leiper N (2008) Partial industrialization of tourism systems: a new model. Curr Issues Tour 11(3):207–235

Lemelin RH, McIntyre N, Koster R, Johnston M (2010a) Climate disruption and the changing dynamics of polar bear – human interaction in Northern Ontario: a case study of polar bear management in Polar Bear Provincial Park, Ontario, Canada. In: Hall CM, Saarinen J (eds) Tourism and change in polar regions: climate, environment and experience. Routledge, London, pp 104–118

Lemelin RH, Dawson J, Stewart EJ, Maher P, Lück M (2010b) Last chance tourism: boom, doom, and gloom. Curr Issues Tour 13(5):477–493

Löytynoia T (2008) The development of specific locations into tourist attractions: cases from Northern Europe. Fennia 186(1):15–29

Lück M, Maher PT, Stewart EJ (eds) (2010) Cruise tourism in the polar regions: promoting environmental and social sustainability? Earthscan, London

Lundmark L (2006) Restructuring and employment change in sparsely populated areas: examples from Northern Sweden and Finland. Kulturgeografiska institutionen, Umeå

Maher PT (ed) (2007) Arctic tourism (special issues). Polar Geogr 30(1&2)

Maher PT (2011) Antarctic human dimensions: ambassadors for the experience: perspectives from the Ross Sea region. In: Maher PT, Stewart EJ, Lück M (eds) Polar tourism: human, environmental and governance dimensions. Cognizant Communication Corporation, New York, pp 121–141

Maher P, Stewart E (2007) Polar tourism (special issue). Tour Mar Environ 4(2&3)

Maher PT, Stewart EJ, Lück M (2011a) An introduction to polar tourism: human, environmental and governance dimensions. In: Maher PT, Stewart EJ, Lück M (eds) Polar tourism: human, environmental and governance dimensions. Cognizant Communication Corporation, New York, pp 3–13

Maher PT, Stewart EJ, Lück M (eds) (2011b) Polar tourism: human, environmental and governance dimensions. Cognizant Communication Corporation, New York

Malcolm CD, Penner HC (2011) Behavior of Belugas in the presence of whale-watching vessels in Churchill, Manitoba and recommendations for local Beluga-watching activities. In: Maher PT, Stewart EJ, Lück M (eds) Polar tourism: human, environmental and governance dimensions. Cognizant Communication Corporation, New York, pp 54–79

Müller DK (2011) Tourism development in Europe's "last wilderness": an assessment of nature-based tourism in Swedish Lapland. In: Grenier AA, Müller DK (eds) Polar tourism: a tool for regional development. Presse de l'Universite de Quebec, Montreal, pp 129–153

Müller DK, Jansson B (2007) The difficult business of making pleasure peripheries prosperous: perspectives on space, place and environment. In: Müller DK, Jansson B (eds) Tourism in peripheries: perspectives from the Far North and South. CABI, Wallingford, pp 3–18

Roura R (2010) Cultural heritage tourism in Antarctica and Svalbard: patterns, impacts, and policies. In: Hall CM, Saarinen J (eds) Tourism and change in polar regions: climate, environment and experience. Routledge, London, pp 180–203

Ryall A, Schimanski J, Howlid Wærp H (eds) (2010) Arctic discourses. Cambridge Scholars, Newcastle upon Tyne

Saarinen J (2007) Tourism in peripheries: the role of tourism in regional development in northern Finland. In: Müller DK, Jansson B (eds) Tourism in peripheries: perspectives from the far north and south. CABI, Wallingford, pp 41–52

Smith SLJ (1998) Tourism as an industry: debates and concepts. In: Ioannides D, Debbage KG (eds) The economic geography of the tourist industry: a supply-side analysis. Routledge, London, pp 31–52

Snyder JM, Stonehouse B (eds) (2007) Prospects for polar tourism. CAB International, Wallingford

Stewart EJ, Draper D, Johnston ME (2005) A review of tourism research in polar regions. Arctic 58(4):383–394

Stonehouse B, Snyder JM (2010) Polar tourism: an environmental perspective. Channel View, Bristol

Tervo-Kankare K, Saarinen J (2011) Climate change and adaptation strategies of the tourism industry in northern Europe. In: Maher PT, Stewart EJ, Lück M (eds) Polar tourism: human, environmental and governance dimensions. Cognizant Communications Corporation, New York, pp 221–235

Timothy DJ (2010) Contested place and legitimization of sovereignty through tourism in polar regions. In: Hall CM, Saarinen J (eds) Tourism and change in polar regions: climate, environment and experience. Routledge, London, pp 288–300
UNWTO (2012) World tourism barometer, vol 12, March. UNWTO, Madrid
van Polanen Petel T (2011) A case study on the effects of human activity on Weddell Seals (Leptonychotes weddelli) with management outcomes. In: Maher PT, Stewart EJ, Lück M (eds) Polar tourism: human, environmental and governance dimensions. Cognizant Communication Corporation, New York, pp 142–166
Wall Reinius S (2009) A ticket to national parks? Tourism, railways and the establishment of national parks in Sweden. In: Frost W, Hall CM (eds) Tourism and national parks: international perspectives on development, histories and change. Routledge, London, pp 184–196

Chapter 2
Looking Back, Venturing Forward: Challenges for Academia, Community, and Industry in Polar Tourism Research

Patrick T. Maher

Abstract This chapter will address questions of what polar tourism is and how it has been researched. These questions serve to allow the field to look back upon its research and reflect. However, looking back is not enough. Within a group such as the International Polar Tourism Research Network (IPRTN), we must also attempt to venture forward and ask: how should polar tourism be researched? On the surface these questions seem simple, but when one dives deeper, there is considerable complexity to consider. This chapter will put such questions in context by offering expert reflection on them from three perspectives: academic, community, and industry. This will be done within a geographical scope of both poles and across decades of work. While one can never expect to capture absolutely everything about these questions, this chapter will hopefully contribute to an ongoing conversation of research practices and research relevance moving into the future.

Keywords Polar tourism • Research • Past • Present • Future

2.1 Introduction

To begin this chapter there are two critical questions that must be addressed: what is polar tourism and how has it been researched? Further to this is a follow-up question that is worth asking: how should polar tourism be researched? These are not easy questions to address, but they are of intimate concern for a group such as the International Polar Tourism Research Network (IPTRN). This chapter will examine such questions through the lens of expert reflection (as the author gave the 2nd IPTRN conference keynote address) versus specific empirical work. A broad

P.T. Maher, Ph.D. (✉)
Outdoor Recreation and Tourism Management, University of Northern British Columbia, 3333 University Way, Prince George, BC V2N 4Z9, Canada
e-mail: maherp@unbc.ca

D.K. Müller et al. (eds.), *New Issues in Polar Tourism: Communities, Environments, Politics*, DOI 10.1007/978-94-007-5884-1_2,

net has been cast to collect information for reflection on these questions: crossing academic, community, and industry lines, from one pole to the other, and across decades of work. However, one can never expect to capture absolutely everything about these questions, this chapter will simply add to the discourse, or perhaps begin it for some readers.

The question of polar tourism research could be solely contained within the academic realm, but that would not be constructive. Tourism is an industry that interacts with a variety of attractions, one of which is the diverse set of communities across the geographical north and the disciplinary south. Thus, this chapter will examine polar tourism research from the three perspectives of academia, industry, and community but also reveal changes that have occurred over time.

Back to the initial question, what is polar tourism? Simply put, polar tourism is tourism in and about the polar regions (Maher 2008a; Maher et al. 2011). Delineating Antarctica is relatively easy; land and sea south of 60°S is considered Antarctica, as outlined by the Antarctic Treaty System ("Antarctic Treaty" 2007). Subsequently, the sub-Antarctic is land and sea between about 60°S and 45°S, including all the islands of the Southern Ocean and the tip of South America (Maher et al. 2011) (see Fig. 2.1).

However, when one uses a geographical boundary in the Arctic, there is the confusion that "polar bears and tundra are found at 51°N in eastern Canada and agriculture is practiced beyond 69°N in Norway" (Nuttall 2005, p. 121). Other boundaries could also be used: political boundaries, such as the territories and portion of provinces north of 60°N in Canada or the Arctic Circle; or biophysical boundaries, such as the tree line or the July 10°S isotherm (see Maher et al. 2011). Maher (2007a) suggests that because large parts of the Arctic are dependent on the marine environment, a marine delineation of the Arctic would be useful and appropriate. It is the marine component that produces the examples used by Nuttall (2005). To make matters even more confusing, there are also different terms used interchangeably by different authorities, such as the following: "North," which in the Canadian context could mean everything from Baffin Island to Ontario's Muskoka cottage country because of perspective; "High North," a term used mainly in Norway; "Circumpolar North"; and "the Arctic." The Arctic Human Development Report defines Arctic as encompassing "all of Alaska, Canada north of 60°N together with northern Québec [Nunavik] and Labrador [Nunatsiavut], all of Greenland, the Faroe Islands, and Iceland and the northernmost counties of Norway, Sweden and Finland... [in Russia] the Murmansk Oblast, the Nenets, Yamalo-Nenets, Taimyr, and Chukotka autonomous okrugs, Vorkuta City in the Komi Republic, Norilsk and Igsrka in Krasnoyarsky Kray, and those parts of the Sakha Republic whose boundaries lie closest to the Arctic Circle" (Stefansson Arctic Institute 2004, pp. 17–18) (Fig. 2.2).

Placing geographical definitions aside, the term polar could also refer to an area's remoteness, which is the essence of Hamelin's (1978) notion of "nordicity," and within tourism this sociological perspective receives attention by Grenier (1998, 2004). Therefore, polar tourism in and about the polar regions could include visits to museums that focus on polar topics (i.e., the International Antarctic Centre in

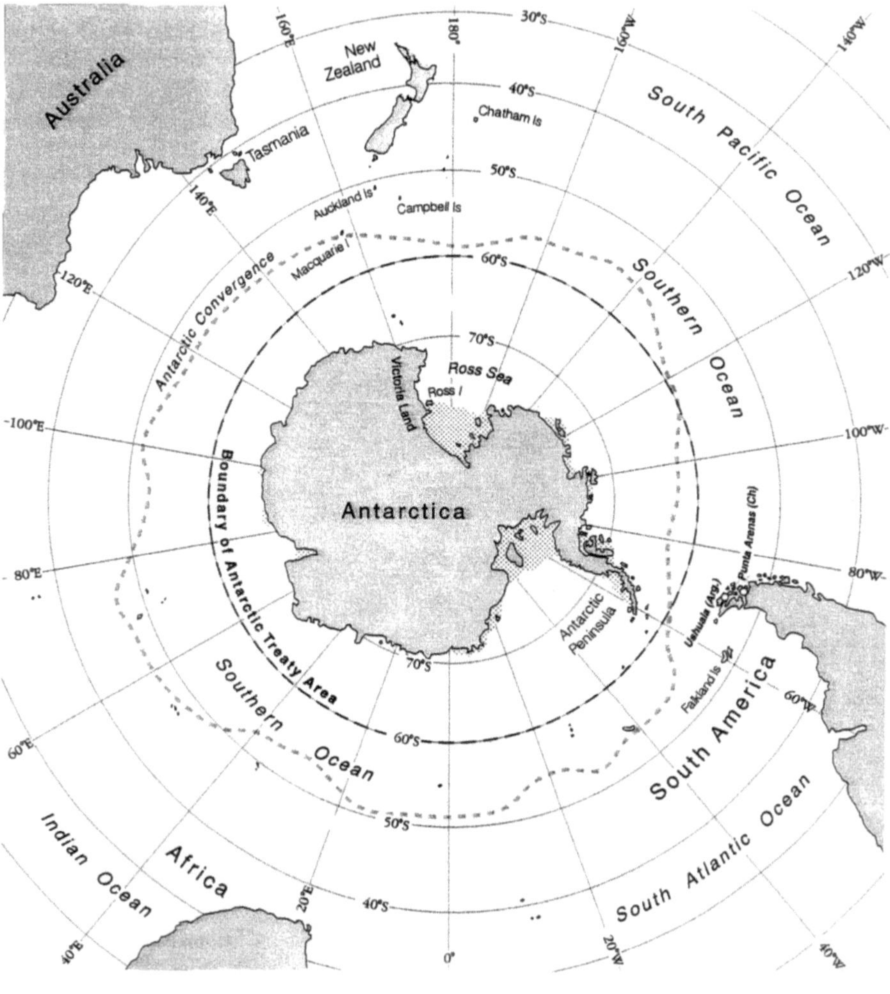

Fig. 2.1 The Antarctic – a regional map (Source: New Zealand Department of Conservation)

Christchurch, NZ, or, in a research sense, the Scott Polar Research Institute in Cambridge, UK) or taking part in activities primarily attributed to the polar regions (i.e., soap stone carving, as practiced by the Inuit people, but that could be learned at workshops of the North House Folk School in southern Grand Marais, Minnesota).

With the where and what of polar tourism discussed, what is there to research? Why would one care to research the subject? This answer lies in the interaction of tourism numbers, on a fragile environment, where values placed upon items such as wilderness, culture, and science run up against the business or economic entity that is tourism. Recent figures show polar tourism to be quite large; Table 2.1 shows some of the recent estimates.

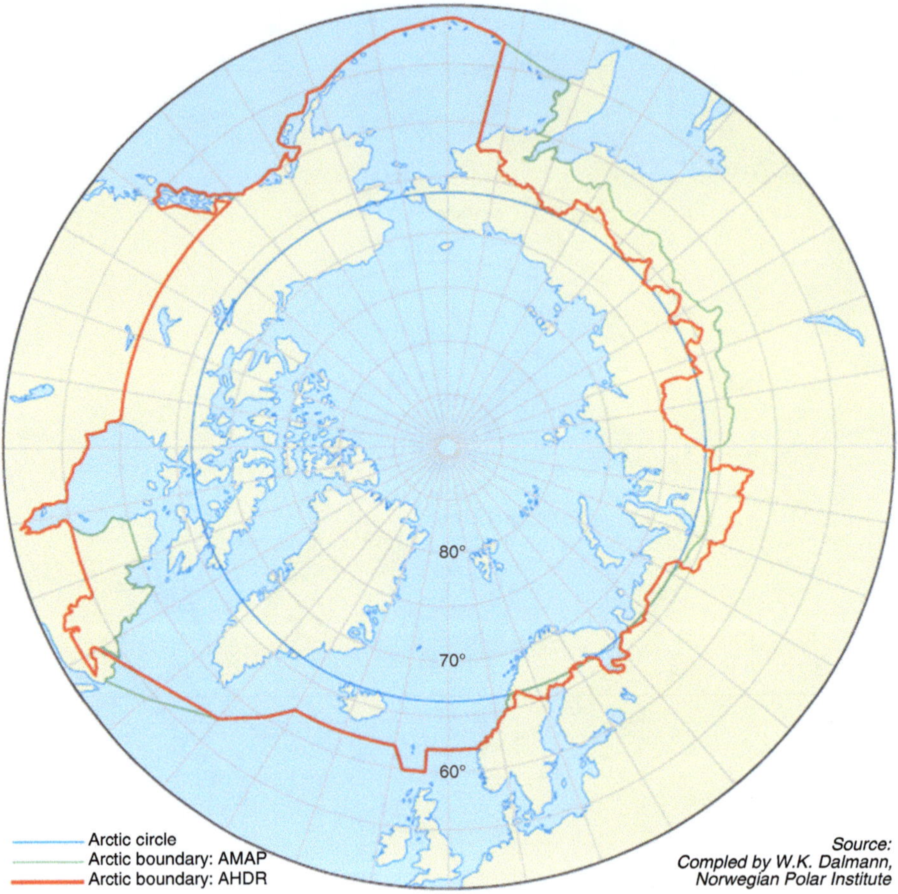

Fig. 2.2 The Arctic – a regional map (Source: Arctic Human Development Report, 2004)

So with these sorts of numbers being present today, research and examination of the industry has become quite important for academia, the industry itself, and the local communities that are influenced by this tourism. The next three sections will examine where that research examination has come from, where it is today, and where it is headed.

2.2 Polar Tourism Research

This chapter, and the presentation it is based upon, grew out of the intense attention the polar regions received in the past due to the International Polar Year (IPY: 2007–2009). Yet surprisingly little research examined tourism despite the tremendous

Table 2.1 Estimates of tourist numbers to a variety of polar region destinations. Modified and updated by the author from Dawson et al. (2009) and Lück et al. (2010)

Country/region/province	Tourist numbers (estimates)	Sources/notes
Canada		
Northwest Territories	33,000–40,000	Lemelin and Johnston (2008)
	62,045	Northwest Territories Industry, Tourism and Investment (2007); 2006–2007 data for all non-resident travellers to the entire territory
Yukon	8,049	Yukon Department of Tourism and Culture (2006); 2004 data – covers only the Northern Yukon tourism region
	112, 891	Yukon Department of Tourism and Culture (2009); Motorcoach visitors making border crossings from cruises docking in Skagway, Alaska offering add-ons
Nunavut	13,000	Buhasz (2006)
	33,000	Milne (2006)
	9,323	Datapath (2006); summer only
Northern Manitoba	2,100–3,000	Lemelin (2005); specific visitors to the Churchill Wildlife Management Area
	6,300–9,000	Lemelin and Smale (2007); actual visits into the CWMA
Nunavik (Northern Quebec)	25,000	Tourisme Quebec, (2004); in 2004, Nord du Quebec statistics included both the Nunavik and James Bay regions
Nunatsiavut (Northern Labrador)	565	Maher and Lemelin (2011); 2008 visitors to Torngat Mountains National Park
	21,000	Hull (2001); adventure tourists to all of Labrador
USA (Alaska)	1,631,500	State of Alaska (2007); summer 2006 data for all out-of-state visitors
	1,026,600	McDowell Group (2010); cruise visitors in 2009, including those who exit Alaska by air following their cruise
Greenland	33,000 (air arrivals) 22,051 (cruise arrivals)	Kaae and Råhede (2011)

(continued)

Table 2.1 (continued)

Country/region/province	Tourist numbers (estimates)	Sources/notes
Iceland	277,800	Lemelin and Johnston (2008)
	69,000	Icelandic Tourist Board (2010); 2009 visitors to Reykjavik aboard 80 vessels
Svalbard (Norway)	8,459	AECO personal communication, August 2010; 2009 expedition cruise visitors (those touring around Svalbard)
	29,813	AECO personal communication, August 2010; 2009 cruise visitors arriving from overseas
Norway (Finnmark)	2,420,959	Viken (2008); data from 2002
Sweden (Norrbotten county)	1,700,000	Zillinger (2007); data from 2001 tourist overnight stays
Finland (Finnish Lapland)	2,117, 000	Hakkarainen and Tulentie (2008); 2006 data for the number of registered tourist overnights
Russia	Estimated at a few tens of thousands and growing steadily – data difficult to obtain	See Pisarev (2008) and Tsekina (2008) for general descriptions
Antarctica	33,824	IAATO (2011)

growth of tourism in the polar regions, particularly the Antarctic, and given the focus in previous IPYs on geographical exploration, which has to some degree become a precursor to tourism (Maher 2007a).

Much research, across both hemispheres, was concentrated at the poles for these 2 years, but again despite the increased research attention, tourism research was largely left out of projects officially linked to the IPY (see www.ipy.org). Again this was surprising given that there are so many issues to which tourism is linked, and thus, tourism research could prove useful; i.e., tourism is seen as a key economic diversification tool for small communities throughout the Arctic, and access for and impacts from tourist vessels often enter the climate change debate surrounding both poles.

2.3 Looking Back

2.3.1 Academia

While largely left out of the IPY, this was not because of lack of research occurring before 2007. As early as the mid-1970s, academics had begun to study the development of tourism in both peripheries (see Reich 1979, 1980; Keller 1982, 1987).

Beginning in the early 1990s, a number of important texts were published and helped to provide a base for the growing interest in polar tourism research today. Three initial texts resulted from conferences (Kempf and Girard 1992; Johnston and Haider 1993; Martin and Tyler 1995), while a fourth was the 1994 special issue on Antarctic tourism in the *Annals of Tourism Research* (1994). Hall and Johnston's (1995) *Polar Tourism: Tourism in the Arctic and Antarctic Regions*, although now out of print, was the first comprehensive overview of tourism issues across both polar regions, addressing issues such as monitoring tourism impacts, the regulation of tourism, patterns of tourism, the impact on aboriginal peoples, issues of access, and the search for sustainable management regimes (Maher and Stewart 2007).

Other contributions followed, including Johnston et al.'s (1998) *Shaping tomorrow's North*; Bauer's (2001) *Tourism in the Antarctic: Opportunities, Constraints and Future Prospects*; Sahlberg's (2001) *Going North: Peripheral Tourism in Canada and Sweden*; and Watson et al.'s (2002) *Wilderness in the circumpolar North: Searching for compatibility in ecological, traditional and ecotourism values*. There were also numerous book chapters on the subject including Stonehouse's (2001) chapter in the *Encyclopedia of Ecotourism – Polar Environments* and Marsh's (2000) chapter in *Tourism and National Parks – Tourism and National Parks in Polar Regions*.

Since 2000, there have been many significant changes in polar tourism (i.e., diversification among products available, increasing tourist numbers, increasing local involvement, and recognition of the importance of local involvement, changes in policy, and governance) (Maher and Stewart 2007). These changes are in addition to ongoing issues such as cumulative environmental and social impacts, provision of education and awareness for tourists (creating ambassadors), increasing vessel sizes and numbers within the cruise sector, and changing climate patterns that affect actual travel as well as the industry (Maher and Stewart 2007).

These more recent research issues have been addressed sporadically in individual book chapters (Bauer and Dowling 2006; Maher 2007b) but also in three edited books published just before the IPY: Baldacchino's (2006) *Extreme Tourism: Lessons from the World's Cold Water Islands*, Müller and Jansson's (2007) *Tourism in Peripheries*, and Snyder and Stonehouse's (2007) *Prospects for Polar Tourism*. Each text had some useful coverage from a research perspective, but was not without their flaws. Baldacchino's (2006) text offers mainly descriptive case study material to share the situation of tourism in cold water locations (from both the northern and southern polar regions) with very little empirical research reported upon. Müller and Jansson's (2007) book addresses polar and subpolar regions in some chapters, but the majority of chapters are not directly dealing with polar tourism rather tourism in periphery locations (whether polar or not). Snyder and Stonehouse's (2007) text is heavily focused on tourism in Antarctica and in general is also much more descriptive than empirical.

2.3.2 Industry

Before, the IPY industry was often very much relied upon for research. Although not an active partner to academia at this stage, industry did monitor itself for impacts and assisted on projects as desired. Project Antarctic Conservation (as discussed by Stonehouse and Crosbie 2007) is a good example of industry/academia cooperation. Industry likely started to receive more and more requests for collaboration in research as researchers themselves began to seek applied topics (which would obviously be of use to industry) versus the traditional "blue sky" research approach. The reliance on industry and their use for research content sometimes led to industry having "a bad taste in their mouths," when they were asked for so much, and the results did not seem to justify the means. Industry rightly became skeptical of the academic community.

Industry also began to materialize itself in research in the early 1990s with associations such as the International Association of Antarctica Tour Operators (IAATO) and later the Association of Arctic Expedition Cruise Operators (AECO). These associations started small, but now include most of the industry players and are the go to source for academics when it comes to industry statistics, an important piece of most research.

2.3.3 Community

In the past community interaction with research was largely limited. Communities and the situation of tourism development were "studied" (Keller 1982). Researchers seemed to come and go; they studied what they needed and then left with very little feedback loop. As with industry, this became increasingly frustrating to communities. In the Antarctic, there was virtually no Antarctic tourism research community (Reich 1979, 1980), and the community interaction was the definitive split between most social science researchers (tourism included) from the science community. The exception here is the very early work by Stonehouse (1965a, b), which covered tourism while having a primary purpose of studying penguin ecology.

2.4 Today

Maher and Gelter (2010) report that one of the most useful developments for polar tourism research during the International Polar Year (IPY) is the increased opportunity for all groups involved in the industry (researchers, students, community members, operators, etc.) to meet and network. This level of coordination (whether overtly planned or not) had not happened since the early 1990s except in the instances of the annual operator meetings of groups such as AECO and IAATO, which on occasion include academic researchers and the odd community member.

2.4.1 Academia

Rising from a gap in publications showcasing empirical work before the IPY, two more additions to the base of polar tourism literature were published early in the IPY: special journal issues of *Polar Geography* (2007) and *Tourism in Marine Environments* (2007). These provided a depth of empirical work that otherwise had not been published together in comprehensive volumes. The Polar Geography special issue was Arctic specific and arouse from the 2006 Canadian Association of Geographers (CAG) annual meeting in Thunder Bay, Ontario, Canada. The *Tourism in Marine Environments* special issue began in response to the distinct lack of tourism-related projects officially linked to the IPY and then moved toward a means to share the work of empirical authors versus descriptive authors. This issue sought to bridge the gap having early career academics publish together and allowing established academics to review such works – creating a bridge between the foundations of the field and the wave moving it forward today. The overwhelming response to the call for this issue then led the editors to successfully solicit for an empirically based book (Maher et al. 2011).

One of the few tourism projects officially linked to the IPY was the *Tourism and Global Change in Polar Regions* conference held in 2007 in Oulu, Finland. The conference was organized jointly by the International Geographical Union's Commission on the Geography of Tourism, Leisure, and Global Change and the Thule Institute at the University of Oulu. As Maher and Gelter (2010) note, this conference provided an excellent opportunity for a broad spectrum of polar tourism researchers to network early on in the IPY and share their recent research findings. Outcomes of this conference include the proceedings (Saarinen and Tervo 2008), another journal special issue (Hall and Saarinen 2008), and a book (Hall and Saarinen 2010).

After Oulu, the next two conferences were held as initial meetings of the two newly developed groups: the University of the Arctic Thematic Network on Northern Tourism (UATNNT) and the IPTRN. The UATNNT (see http://www.uarctic.org/) was established by a working group of researchers from tertiary institutions across the circumpolar north. The institutions that each researcher belonged to are members of the virtual University of the Arctic (see http://www.uarctic.org), and thus, delegates were individually invited to Gjesvaer, Norway, first and foremost as an opportunity to share their personal interests in polar tourism but also because of their institutions' support for the field of tourism. From this starting point the group sought to envision how these multiple interests and levels of support might fit together under the umbrella of a thematic network and in turn support a graduate program of some description.

Sletvold and Maher (2008) presented the group's declaration from the working group meeting in Gjesvaer, Norway, along with a proposal for a joint master's program to the University of the Arctic's Council meeting in Edmonton, Alberta, Canada in June 2008. The network was approved in Edmonton and had funds to operate for the next 2 years, continuing master's program discussions among other topics. The IPTRN was initiated as a second outcome to the polar tourism sessions held at the 2006 Canadian Association of Geographer's conference in Thunder Bay, the first outcome being the 2007 special issue of the journal *Polar Geography*. Following 2 years of initial set up, the IPTRN held its first meeting in Kangiqsujuaq, Nunavik, in August 2008 (www.polartourismnetwork.uqam.ca). While the cost of attending a conference in the remote community of Kangiqsujuaq was a limiting factor, it did allow for increased community involvement (although community delegates came and went).

As a result of this initial meeting, the IPTRN has been founded as a group with a shared interest in research that advances the understanding of tourism in and about the polar regions. The IPTRN strives to generate, share, and disseminate knowledge, resources, and perspectives on polar tourism and strongly supports the development of international collaboration and cooperative relationships between members. Membership in the network can include individuals such as university researchers, consultants, tourism operators, government organizations, community members, and graduate students. The outcome of the first IPTRN conference is the book by Grenier and Müller (2011).

The Vienna Symposium on polar tourism was initiated by a single researcher's desire to better link his own research in the high altitudes (alpine landscapes) with the IPY and his perception that there must be research occurring at high latitudes as well. This self-initiation, generally outside of the circles of the other three conferences, actually led the Vienna Symposium to be quiet unique. While the depth of polar tourism knowledge presented was not the same as that found in the other instances, the Vienna Symposium had more breadth in connecting to

Russian researchers (largely due to Austria's historic connections). An outcome of the symposium is the proceedings (Breiling 2008) available online. The Vienna Symposium also produced a declaration of the discussion held there (Maher 2008b), which was subsequently presented to European funding and research agencies. Within the declaration the themes of education, networking, interdisciplinarity, and impacts were identified along with some important information needs.

Wider than simply a focus on tourism, four other networks saw increased activity among tourism researchers during the IPY as locations to share their work. The Social Sciences and Humanities Antarctic Research Exchange (SHARE; see www.share-antarctica.org) hoped to build a collaborative platform similar to the Scientific Committee on Antarctic Research (SCAR), but first the community needed to mobilize social science and humanities research knowledge with a focus on the Antarctic and thus allow for its use to be seen by governments and policy-makers (Tourism in Marine Environments 2007, p. 245). A similar northern initiative was the Northern Research Network (see http://northernresearchnetwork.electrified.ca). Additionally, given the IPY focus on promoting the emergence of young researchers, the Association of Polar Early Career Scientists (APECS; see http://arcticportal.org/apecs) began playing an important role for a number of researchers and through topic-specific working groups so too did the Ocean Management Research Network (OMRN; see http://www.omrn-rrgo.ca). Even well-established organizations such as the International Arctic Social Science Association (IASSA; see www.iassa.gl) have seen an increased discussion of tourism as witnessed by sessions at their latest meetings in Greenland (during the IPY) and afterward in Iceland.

2.4.2 *Industry/Community*

During the IPY both the industry and community saw themselves very much reengaged. Communities in the Arctic became partners in a wide variety of research, the scientific community in the Antarctic saw numerous new projects evolve, and industry became engaged as media attention caused large populations to want to visit. Did either community or industry become overwhelmed or overstudied? This cannot be immediately evident for all the research from IPY, but given the dearth of tourism research, it is unlikely. Moving ahead, tourism research appears to be on the minds of industry and community as a field that needs some important consideration.

2.5 Venturing Forward

2.5.1 *Academia*

As academia moves forward, there are four areas to work on: celebrate new ideas, create collaboration, update publications, and cooperate. New ideas could be new technologies for research, new methods, or better combination of methods. Polar tourism, and tourism in general, is theory poor, so there need to be allowances for better theoretical linkages or theory building. All of these ideas lead to more holistic projects.

Creating collaboration includes connecting with broader conferences such as 2010's Oslo Science Conference (http://ipy-osc.no). Tourism sessions held in Oslo were well attended and allowed for connection to the wider scientific community. Two publications (a journal special issue in *Polar Research* and a text with Springer) are already underway to allow for this. Inter-conference and intergroup collaboration is also well underway as a researcher could also attend the third UATNNT meeting that was held in Svolvær, Norway, directly after the Oslo Science Conference and then also directly before the second IPTRN conference in Abisko, Sweden. Collaboration also includes attempts at creating circumpolar research projects such as Norway's Chair in Arctic tourism research consortium (http://www.hifm.no/nor/www_hifm_no/hogskolen-i-finnmark-_-startside/arctic-chair/arcticchair?lang=nor). In addition, the continuation of current networks, continued growth, and revision will be important. The UATNNT has now developed three master's level courses and is awaiting approval to begin publicity and active enrollment, and SHARE network has now morphed into the Social Science Action Group (SSAG) of SCAR (see http://www.scar.org/researchgroups/via/).

Since the IPY, there has been the two conference books already mentioned (Grenier and Müller 2011; Hall and Saarinen 2010), and an additional four other new polar tourism texts (and special issues) have become available. These include TÉOROS (2009), Lück et al. (2010), Stonehouse and Snyder (2010), and Maher et al. (2011).

Cooperation should manifest itself through the loss of politics. Researchers need to work together, with each other, with industry, and with wider communities (whether actual or theoretical). The creation of synergies versus isolationism needs to prevail, and engagement with the full suite of partners can help this materialize.

2.5.2 Industry

The tourism industry in the polar regions still sees some abuse of their logistical capacity and researcher desire to use them for their visitor access and capacity without full engagement. There is lingering disengagement from the past, and researchers who seek to continue in this field must recognize the wrongs of the past. With the industry associations, there are growing partnerships, but these are delicate relationships often built over time and with specific results desired. Industry has a renewed enthusiasm for the work of tourism researchers, but it should be cultivated cautiously.

2.5.3 *Community*

Moving forward communities should become more and more engaged. The IPY set a nice example of how communities can lead the way – ask the questions they need answers to and hold researchers accountable to provide feedback and results. The movement to community-based or citizen science will provide "real" feedback, as communities will expect to be engaged every step of the way. In the Antarctic, the community aspect will be better connections between the scientific community and tourism researchers. For many years tourism researchers and indeed most social scientists perceived that they were "on the outside" of the community. The IPY may have allowed many social scientists in, but perhaps more so it was the formation of SHARE and then SSAG.

2.6 Conclusions

To conclude, all three segments of the sphere need an understanding of where the state of polar tourism research is. Academia cannot be ambivalent of industry practicalities. Neither can discount the needs of communities (in the Arctic), and industry and tourism academics must connect to the rest of the scientific community in the Antarctic. Once everyone is "at the table," then the discussion of how to proceed can occur.

Everyone needs to be involved (academics, consultants, community members, government agencies, industry) across a wide breadth of disciplinary perspectives. Through this a research agenda/strategy could create an action plan, provided it were able to receive appropriate funding. If such a strategy were a larger overarching puzzle to solve, then the many smaller pieces could be pursued by individual groups and funded by specific nations/programs. Any puzzle needs to be broad to account for geographical and disciplinary differences.

Research agendas have been generally discussed at the most recent meetings of both the UATNNT and IPTRN; specific research questions were asked at the 2006 CAG (see Maher 2007a) and in Snyder's (2007) UNEP report, but even more specifically articles titled as research agendas have been published such as Stewart et al. (2005), Mason and Legg (1999).

These agendas set a bar, albeit a disjointed one, and as polar tourism researchers and allies, partners, and stakeholders, the talk needs to stop and move to action. An agenda/strategy needs to be comprehensive enough to cover the large-scale topics of the day (i.e., climate change), but open enough to be broken down into manageable pieces by the individual researchers/research consortiums. As discussed at the Vienna Symposium (Maher 2008b), researchers need to immerse themselves in the wide range of empirical work available across disciplines and build from there; if this cannot occur, then the field is doomed to continue describing "what" is happening, "when," and by "whom" and never furthering knowledge of "why" or using such knowledge to appropriately detail "how" to manage or shape the changing industry into the future.

The IPTRN is, in my opinion, the ideal organization to initiate this action on a research agenda/strategy with the full range of appropriate partners. Building on the research found throughout this book, showcasing the work presented at the 2nd Conference of the International Polar Tourism Research Network that examined Tourism, People, and Protected Areas in Polar Wilderness, we can gain a lot and make headway in Nain, Nunatsiavut in 2012. To what degree tourism research is included in a future IPY will largely depend on taking up such a call. Thus, it is with optimism that the titles of the next IPY follow-up conference in Montreal (2012) and the 3rd IPTRN conference (2012) both have action in their titles, respectively: From Knowledge to Action and From Talk to Action.

References

Annals of Tourism Research. (1994). Antarctic tourism (Special issue). 21(2)

Antarctic Treaty. (2007). United States of America: National Science Foundation, Office of Polar Programs, January 26. Retrieved 14 March 2007 from, http://www.nsf.gov/od/opp/antarct/anttrty.htm

Baldacchino G (ed) (2006) Extreme tourism: lessons from the world's cold water islands. Elsevier, Amsterdam

Bauer TG (2001) Tourism in the Antarctic: opportunities, constraints and future prospects. The Haworth Hospitality Press, New York

Bauer TG, Dowling RK (2006) The Antarctic cruise industry. In: Dowling RK (ed) Cruise ship tourism. CABI, Oxon, pp 195–205

Breiling M (ed) (2008) The Vienna symposium on polar tourism. Vienna University of Technology – Technology, Tourism and Landscape Centre, Vienna, Austria. Retrieved from http://ttl.tuwien.ac.at/docs/proceedings_polar_tourism_081021.pdf

Buhasz L (2006) Northern underexposure. Globe & Mail, July 1

Datapath (2006) Nunavut exit survey. Datapath Consulting, Marsh Lake

Dawson J, Stewart EJ, Maher PT, Slocombe DS (2009) Climate change, complexity and cruising in Canada's Arctic: a Nunavut case study. In: Anderson R, Bone RM (eds) Natural resources and aboriginal people in Canada, 2nd edn. Captus Press Inc, Concord

Grenier AA (1998) Ship-based polar tourism in the northeast passage: a case study. Faculty of Social Sciences, University of Lapland, Rovaniemi

Grenier AA (2004) The nature of nature tourism. PhD thesis, Faculty of Social Sciences, Acta Universitatis Lapponiensis 74. University of Lapland, Rovaniemi

Grenier A, Müller DK (eds) (2011) Polar tourism: a tool for regional development. UQAM Press, Montreal

Hakkarainen M, Tulentie S (2008) Tourism's role in rural development of Finnish Lapland: interpreting national and regional strategy documents. Fennia 186(1):3–13

Hall CM, Johnston ME (eds) (1995) Polar tourism: tourism in the Arctic and Antarctic regions. Wiley, Chichester

Hall CM, Saarinen J (2008) Tourism and change in northern geographies. Fennia Int J Geogr 186(1 [Special issue])

Hall CM, Saarinen J (eds) (2010) Tourism and change in the polar regions: climate, environments and experiences. Routledge, London

Hamelin L-E (1978) Canadian nordicity: it's your north too (trans: Barr W). Harvest House, Montreal

Hull JS (2001) Opening up the big land to the world: the role of the public sector in adventure tourism development in Labrador. In: Sahlberg B (ed) Going north: peripheral tourism in Canada and Sweden. European Tourism Research Institute, Östersund

IAATO (International Association of Antarctica Tour Operators) (2011) 2010–2011 tourism statistics. Retrieved from http://iaato.org/tourism_stats.html

Icelandic Tourist Board (2010) Tourism in Iceland in figures. Available at http://www.ferdamalastofa.is/upload/files/Tourism_in_Iceland_in_%20figures_Feb2010.pdf

Johnston ME, Haider W (eds) (1993) Communities, resources, and tourism in the north. Lakehead University, Centre for Northern and Regional Studies Series, Thunder Bay

Johnston ME, Twynam GD, Haider W (eds) (1998) Shaping tomorrow's north: the role of tourism and recreation. Lakehead University, Centre for Northern Studies, Thunder Bay

Kaae BC, Råhede MK (2011) Governance dimension: tourism strategies and environmental labelling in Greenland. In: Maher PT, Stewart E, Lück M (eds) Polar tourism: human, environmental and governance dimensions. Cognizant Communication, Elmsford

Keller CP (1982) The development of peripheral tourist destinations: case study, Baffin Region. Unpublished Master of Arts thesis, University of Western Ontario, London, ON, Canada

Keller CP (1987) Stages of peripheral tourism development – Canada's northwest territories. Tour Manage 8(1):20–32

Kempf C, Girard L (eds) (1992) Tourism in polar areas: proceedings of the first international symposium, Colmar, France, 21–23 April 1992. International Union for Conservation of Nature and Natural Resources and Ministry of Tourism (France), Neuilly Sur Seine

Lemelin RH (2005) Wildlife tourism at the edge of chaos: complex interactions between humans and polar bears in Churchill, Manitoba. In: Berkes F, Huebert R, Fast H, Manseau M, Diduck A (eds) Breaking ice: renewable resource and ocean management in the Canadian North. University of Calgary Press, Calgary

Lemelin RH, Johnston ME (2008) Arctic tourism. In: Lück M (ed) The encyclopedia of tourism and recreation in marine environments. CABI, Wallingford, pp 32–33

Lemelin RH, Smale B (2007) Wildlife tourist archetypes: are all polar bear viewers in Churchill, Manitoba ecotourists? Tour Mar Environ 4(2–3):97–111

Lück M, Maher PT, Stewart EJ (eds) (2010) Cruise tourism in polar regions: promoting environmental and social sustainability? Earthscan, London

Maher PT (2007a) Arctic tourism: a complex system of visitors, communities, and environments. Polar Geogr 30(1–2):1–5

Maher PT (2007b) Wildlife and tourism in Antarctica: a unique resource and regime for management. In: Higham J, Lück M (eds) Marine wildlife and tourism management: insights from the natural and social sciences. CABI, Oxford, pp 294–318

Maher PT (2008a) International polar years. In: Lück M (ed) The encyclopedia of tourism and recreation in marine environments. CABI, Wallingford, pp 243–244

Maher PT (2008b) Declaration from the Vienna symposium on polar tourism. Retrieved from http://ttl.tuwien.ac.at/docs/Declaration_of_the_Vienna_Symposium_on_Polar_Tourism.pdf

Maher PT, Gelter H (2010) A commentary on polar tourism conferences in the international polar year (2007–2009). Tour Mar Environ 6(4):185–188

Maher PT, Lemelin RH (2011) Northern exposure: opportunities and challenges for tourism development in Torngat mountains National Park, Labrador, Canada. Polar Rec 47(240): 40–45

Maher PT, Stewart EJ (2007) Polar tourism: research directions for current realities and future possibilities. Tour Mar Environ 4(2–3):65–68

Maher PT, Stewart EJ, Lück M (eds) (2011) Polar tourism: human, environmental and governance dimensions. Cognizant Communications Corp, Elmsford

Marsh J (2000) Tourism and national parks in polar regions. In: Butler RW, Boyd SW (eds) Tourism and national parks: issues and implications. Wiley, Chichester, pp 125–136

Martin VG, Tyler N (eds) (1995) Arctic wilderness: the 5th World Wilderness Congress, Tromsø, Norway, 1993. North American Press, Golden
Mason PA, Legg SJ (1999) Antarctic tourism: activities, impacts, management issues, and a proposed research agenda. Pac Tour Rev 3(1):71–84
McDowell Group (2010) Economic impact of Alaska's visitor industry. McDowell Group, Juneau
Milne S (2006) Baffin Island, Nunavut, Canada. In: Baldacchino G (ed) Extreme tourism: lessons from the world's cold water islands. Elsevier, Oxford
Müller DK, Jansson B (eds) (2007) Tourism in peripheries. CABI, Wallingford
Northwest Territories Industry, Tourism and Investment (2007) Canada's Northwest territories: tourism & parks: tourism research and statistics (Online). http://www.iti.gov.nt.ca/parks/tourism/research_and_statistics.htm. Accessed 28 Sept 2007
Nuttall M (2005) Arctic: definitions and boundaries. In: Nuttall M (ed) Encyclopedia of the Arctic, vol 1. Routledge, New York, pp 117–121
Pisarev SV (2008) Tourism in polar oceans: advantages and disadvantages of cooperations with marine investigations. In: Breiling M (ed) The Vienna symposium on polar tourism; October 22–25 2008, Vienna, Austria. Proceedings. Vienna University of Technology – Technology, Tourism and Landscape Centre, Vienna
Polar Geography (2007) Arctic tourism (Special issue). 30(1–2)
Reich RJ (1979) Tourism in the Antarctic: its present impact and future development. Unpublished diploma in Polar Studies dissertation, Scott Polar Research Institute, Cambridge University, Cambridge, UK
Reich RJ (1980) The development of Antarctic tourism. Polar Rec 20:203–214
Saarinen J, Tervo K (eds) (2008) Tourism and global change in polar regions. University of Oulu Press, Oulu
Sahlberg B (ed) (2001) Going north: peripheral tourism in Canada and Sweden. European Tourism Research Institute (ETOUR), Östersund
Sletvold O, Maher PT (2008) A thematic network on northern tourism. Proposal presentation to the Council of the University of the Arctic – 11th Meeting – June 2–5, 2008, Edmonton, Canada
Snyder J (2007) Tourism in the polar regions: the sustainability challenge. United Nations Environment Program, Paris
Snyder JM, Stonehouse B (eds) (2007) Prospects for polar tourism. CABI, Wallingford
State of Alaska (2007) Alaska visitor statistics program: summer 2006. State of Alaska, Department of Commerce, Community and Economic Development, Juneau
Stefansson Arctic Institute (2004) Arctic human development report. Stefansson Arctic Institute, Akureyri
Stewart EJ, Draper D, Johnston ME (2005) A review of tourism research in the polar regions. Arctic 58(4):383–394
Stonehouse B (1965a) Too many tourists in Antarctica. Animals 7(17):450–453
Stonehouse B (1965b) Counting Antarctic animals. New Scientist, 29 July 1965, 273–276
Stonehouse B (2001) Polar environments. In: Weaver DB (ed) The Encyclopedia of Ecotourism. CABI, Wallingford, pp 219–234
Stonehouse B, Crosbie K (2007) Antarctic tourism research: the first half century. In: Snyder JM, Stonehouse B (eds) Prospects for polar tourism. CABI, Wallingford, pp 210–228
Stonehouse B, Snyder JM (2010) Polar tourism: an environmental perspective. Channel View Publications, Bristol
TÉOROS (2009) Tourisme Polaire (Special issue). 28(1)
Tourism in Marine Environments (2007) Polar tourism (Special issue). 4(2–3)
Tourisme Québec (2004) Tourisme en chiffres 2004 – version finale. Gouvernement du Québec, Québec
Tsekina MV (2008) The usage of the tourist potential of the Northern and Southern Poles. In: Breiling M (ed) The Vienna symposium on polar tourism; October 22–25 2008, Vienna, Austria. Proceedings. Vienna University of Technology – Technology, Tourism and Landscape Centre, Vienna

Viken A (2008) The nature of Arctic tourism. Presentation to the initial meeting of the University of the Arctic Thematic Network on Northern Tourism. Gjesvaer, Norway, May 24, 2008

Watson AE, Alessa L, Sproull J (eds) (2002) Wilderness in the circumpolar north: searching for compatibility in ecological, traditional and ecotourism values. Proceedings from a conference held in Anchorage, Alaska, 15–16 May 2001. USDA Forest Service, Rocky Mountain Research Station, Ogden

Yukon Department of Tourism and Culture (2006) 2004 visitor exit survey: Klondike region. Government of Yukon, Whitehorse

Yukon Department of Tourism and Culture (2009) 2009–2010 tourism Yukon situation analysis. Government of Yukon, Whitehorse. Available at http://www.tc.gov.yk.ca/pdf/Situation_Analysis_2009-10.pdf

Zillinger M (2007) Organizing tourism development in peripheral areas: the case of the Subarctic project in Northern Sweden. In: Müller DK, Jansson B (eds) Tourism in peripheries: perspectives from the far north and south. CABI, Wallingford

Chapter 3
What Is Arctic Tourism, and Who Should Define It?

Implication of an Anglo-American Hegemony in Academic Writing

Arvid Viken

Abstract This chapter discusses the Anglo-American hegemony of academic writing about Arctic tourism. This is often reckoned as a periphery and this term is also discussed. It is argued here that this hegemony results in some off accounts. The chapter shows how some of the writing about Arctic tourism in academic books and journals tends to be very inaccurate and general, adding to a perception of these geographical areas as less developed and modern. However, it is also shown that there are research rather indicating the opposite – that the "periphery" is an integrated part of a global hybridity, but also that to break the hegemony is a challenging matter.

Keywords Arctic • Anglo-American • Hegemony • Tourism • Periphery

3.1 Introduction

This chapter discusses academic writing about Arctic tourism. The argument put forward here is that much of what is written about this topic is biased, in one way or another, and that this partly relates to the Anglo-American hegemony within the field, a dominance of authors, editors, journals, and publishers located in English-speaking countries, in particular the UK and the USA. In geography, there has been an encompassing debate concerning this issue (Gregson et al. 2003; Kitchen 2005). A similar hegemony is also registered in the field of tourism geography. Gibson (2008: 413) shows that more than 80% of the articles in the so-called international journals have authors originating from Anglophone countries. Concerning academic tourism literature in general, Dann and Parinello (2009) show a similar pattern,

A. Viken (✉)
Department of Sociology, Political Science and Community Planning, University of Tromsø, 9037 Tromsø, Norway
e-mail: arvid.viken@uit.no

D.K. Müller et al. (eds.), *New Issues in Polar Tourism: Communities, Environments, Politics*, DOI 10.1007/978-94-007-5884-1_3,

though delimited to the leading journal of the field, *Annals of Tourism Research*; 37 of 51 articles had Anglo-American authors and 97% of the quotations were to English-written sources.

Gramsci is a central scholar in the theory of hegemony (cf. Urbinati 1998). His thoughts emerged from analyses of the "Southern Question": the political and cultural differences between north and south in Italy. His standpoint was that these differences were rooted in the political dominance of the north, but he was convinced that the issue should be solved through general national policies and not by treating the south as a special or deviant case (ibid: 373). In coping with the problem, as he saw it, intellectuals should have major position through their role in the constitution of the political consciousness and the human fabric in general. Thus, the discussion of the Anglo-American hegemony in academic writing – it is not only an issue within geography – shows that this stance probably is wrong. There are also a parallel to hegemonies within academia in the publishing sector (Gregson et al. 2003). The northern tourism issue is part of a global "Northern Question" – how the north should be understood. This of course raises the question of how northern knowledge is produced and should be produced, and it relates to questions of how and by whom the north shall be governed.

This chapter discusses hegemonic tendencies in academic writing about Arctic tourism but takes related academic fields as a point of departure. As an activity enacted on the northern rim, most people will see Arctic tourism as tourism in the periphery. Therefore, this is also discussed. In one part, some examples are given about the publishing impacts of such accounts, that they are inaccurate and partly misleading, and partly support stereotype perception of the Arctic. Toward the end, it is discussed if there is any alternative. The sources for the article are basically what is written about Arctic tourism but with an emphasis on recent contributions. The presentation does not pretend to be statistical, giving a representative picture of the situation. The aim is rather to show problems related to this hegemony. The article does not deny the fact that although there is a hegemony, much that is written is of course of great value. This presentation has an insider's point of view – the author is a North-Norwegian academic – and may well have some of the bias that the outsiders are claimed to have. Research is never objective, although one tries to be balanced.

3.2 Hegemony in Academia

Academic hegemony concerns the relationship between knowledge and power. As Foucault and others have shown, this relation is as much about how power constitutes knowledge as the more traditional opposite (Foucault 1980; Hall 2001). In the discussion about hegemony, the fact that power directs knowledge production is particularly central. The power system within international academia is such that what matters is not only what is written and by whom it is written but from where the authors and articles originate, and particularly in what journals or books things are published. Gregson et al. (2003: 8) use the term "writing spaces," which are "hierarchies of power, which position some journals as mattering rather more than others... capturing, controlling and regulating the international (read theoretical and/or conceptual) high ground, its cutting edges." They also refer to a prevailing rhetoric putting the UK and the USA in the center, seeing non-UK/US geographies as the knowledge about the European Other. And further, theory and concepts are produced in the centers; what is produced elsewhere are illustrative cases or empirical evidence or looked upon as deviancies from these general theories. Although also the general theories often are established on the basis of empirical studies made in the centers or national studies in a particular country, they tend not to be reckoned as case studies in the same way as studies made outside the Anglo-American domain. Referring to areas outside, Gregson et al. (2003: 12) claim that "[c]ontemporary Europe figures as an extensive case-study area," and as Aalbers and Rossi (2006) mention, run the risk of having to be revised according to suggestions by Anglo-American referees. Often, authors have to introduce the reader to the geographical context because they are unfamiliar to the referees or expected Anglo-American readers. Thus, these critiques obviously see Anglo-American academia to be ethnocentric. Flyvbjerg (2006) argues that there scarcely exists anything but case studies, since most studies are restricted to a particular location (a given place, region, or country). Consequently, case studies are the very origin of many theories which are today reckoned as generic knowledge. Yet, and still, within the Anglo-American domain, "... geographies of other people and places become marked as Other – exotic, transgressive, extraordinary, and by no means representative...", according to Berg and Kearns (1998: 129). This system places both the researchers of the center and from outside within a hierarchical system. Quoting Gregson et al. (2003), Kitchen (2005: 6) notes that "many non-Anglo-American geographers are caught between their own national traditions and Anglo-American work, a position that implicitly acknowledges and reproduces a peripheral identity." One of the consequences of this, Kitchen (2005: 6) remarks, is that the theoretical production "casts much of the world's geography into silence." There is a socially constructed center-periphery dimension within academia reflecting a hegemonic structure. Illustrations can be found in many fields.

A Finnish-British study from the field of organization theory illustrates the hegemonic process (Meriläinen et al. 2008). The article is about production of a paper based on a collaboration between Finnish and British researchers. The article shows how the Finnish researcher becomes marginalized during the research and publication process. As the authors themselves see it, the first draft (only authored by the Finnish member) was given referee comments that "marginalised the Finnish experience to the status of 'different', a potentially interesting and/or deviant case vis-à-vis an unstated 'normal' case" (ibid: 590). When the research team presented the findings at a conference, the British teammates substituted the Finnish interview material with British, without really changing the content – the referees had then only minor comments. In writing about this process in the aftermath, the Finnish members of the author team describe this as a process of othering. Being an "other" means not to be part of the reference group – the Anglo-American writers.

The reason for replacing Finnish with British data was obviously to make it stand out as more general and valid. Another reason was the fact that the British researchers were unable to read the Finnish material, due to lack of language skill. The case probably also shows that data from an unfamiliar context is not seen as reliable as data from a "known" environment. There is probably a tendency that some writing cultures are more matter of othering than others, for instance, aboriginal writers. And of course, by raising these questions, there is a risk that this chapter adds to othering processes.

One reason for the Anglo-American hegemony in academia is also related to language. The first point about this is that if you do not write in English, you will not be published or you will be read only by a few. Writing in native non-English means a limited audience, and in many fields – for instance, in tourism – there may well be feeble publishing traditions and lack of relevant journals. And since it is like this, less and less is written in non-English languages. The second point is that to be recognized and merited one has to publish internationally, which means in English. Thus, you have to compete with native English-speaking authors, which means authors that are more likely to be perceived as "devoted to theory" and presenting "real" empirical evidence. But of course, there are lots of case studies also from the Anglophone world that appear as such. The general point is that language is tied to culture; cultures of understanding and writing that all in all represent barriers for non-English-speaking scholars in their publishing efforts. It is also a fact that the publishers also tend to be Anglo-American and within the same language culture and that the systems for citation indexes and ranking the journals also underpin this culture.

The above characteristics of academia may be parts of the explanation of why until recently only a few articles and books that address issues related to tourism in the Arctic could be found. Concerning journals, most of them are based in the UK or the USA and far from the polar regions. The academics in the regions, as Gregson et al. (2003) point out, tend to adapt to the Anglo-American writing traditions and write about the "major" or "normal" issues as are conceived of in these circles. If you focus on the Arctic, you risk to be perceived as deviant. Thus, there is even a risk that those doing so are those that do not succeed in coping with the core issues. However, the body of academics writing about Arctic tourism has grown significantly in recent years, and the publishing pattern has slightly been changed. Since year 2000, an Anglophone but Nordic-based journal has been published, within which it is said be easier for writers from this region to be accepted (Müller 2009). Since 2007, a whole series of books have been published on the topic. Despite most authors are southerners, many also originate from the Arctic or are familiar with Arctic living conditions.

3.3 The Rhetoric of the Arctic and Arctic Tourism

The Arctic is a polar region. Basically, there are only two such regions: the Antarctic and the Arctic. However, as has been discussed in other places, there is no clear definition of the "polar regions" or of the Arctic. As Grenier (2007: 58) emphasizes, each academic discipline tends to have its own ways of delimitating the "polar," but the most common one is to treat the polar circles as the lines that demarcate the polar regions, although Johnston (1995) claims that this primarily is a European delineation (for further discussions, cf. Maher 2007; Stewart et al. 2005). This means that within the Arctic there are modern towns such as Rovaniemi and Tromsø, high-tech sites as Thule and Longyearbyen, indigenous peoples and nonindigenous peoples, and so on. From a tourist market point of view, these regions can also be seen as destinations (Snyder 2007). However, the Arctic is not one destination, as the different parts of the area will not be visited on one single tour. It is a huge area covering parts of Alaska, Canada, Greenland, Iceland, Scandinavia, and Russia.

There is no doubt that tourism in the Arctic is different from tourism in other areas. Certainly, the Arctic has its particular locality giving the area a kind of identity. For most people, it is a remote location, but in a world that is said to be compressed into one place, this is as much a mental construct as a reality. However, the low temperatures, the snow, the dark and the light, the phenomena as northern light and midnight sun are facts, and the Arctic uniqueness refer to these and some other extreme physical characteristics. However, the Arctic is huge, also in a north–south perspective. Thus, whereas in December it is totally dark in the middle of the day on Svalbard, you have 3–5 h of light in the Scandinavian Arctic mainlands. In some places the summer temperature very seldom expires 10°C, and other places can be quite warm. These are the hard facts. However, the Arctic is associated with a lot more. It is also a socially constructed area. It is a vast location, not much of a "region", which often is conceived as an entity with common institutions, policies, and identities (Paasi 2003). However, both within and outside the area, "Arctic" is a known denotation. In a seminar with tourism researchers from Rovaniemi in Finland and Alta in Norway (both located in the Arctic) in 2009, one session was devoted to identifying the essence of Arctic tourism. The group of researchers worked in two separate groups coming up with short characteristics of Arctic tourism. One comment was "Is there any difference from other tourism, except for the location?" The other comments can be grouped in three: those related to geophysical facts, those related to the tourist experiences, and those related to the tourism industry:

- Geophysical comments: Remote area tourism, far north, a particular location, cold climate, nature extremes, wilderness, vulnerable nature.
- Experience-related comments: Unique experience, pure nature and silence, nature-based experiences, togetherness of nature and culture, a place to have been.
- Industrial-oriented comments: Small-scale, not very profitable, publicly funded, regional development, a new way of colonizing the north.

It should be said that this group was limited, for instance, including only a couple of aboriginal researchers. The issue could have been expressed differently with another composition of the group.

Many of terms used by the scholars relate to extremes and to terms that are perceived as one side of a dichotomy or an end of a scale; cold is an antonym of warm and so on. Some of the dichotomies or scales tied to the Arctic are as follows:

Cold	<-------->	Warm
Marginal	<-------->	Core
Remote	<-------->	Close
Polar	<-------->	Nonpolar
Wilderness	<-------->	Inhabited
Rural	<-------->	Urban
Primitive	<-------->	Civilization
Un(der)developed	<-------->	Developed
Periphery	<-------->	Central

The interpretation of the terms, dichotomies, and scales is not arbitrary. First, it is not incidental which terms are put left or right on the scale: most people would put center or core to the left, periphery and margin to the right. This influences the way the reader or listener organizes the information: center first, periphery second. Next, some terms refer to "normality" (center, urban, developed); periphery is defined in relation to center or core, not the opposite. Most terms in the list above are in fact not neutral, but relate to a series of connotations and meanings, often filled with power relations. This way of organizing knowledge often involves a judgment, something is better than something else. When "Arctic" is tied to such terms, dichotomies, and scales, the area is valued normatively and politically and often judged according to standards created by others. To state that the Arctic is unique is rather neutral; to categorize it as marginal, wild, and peripheral is to make normative judgments (cf. Heller 1999). How things are ordered, and what scales are used, is a normative concern. As a former Sami president once declared, the Sami areas are not wild. His people probably have lived there for thousands of years. To him, it is the modern city life that is wild.

The way in which the Arctic is presented reflects existing discourses. A discourse is something that characterizes the thinking about a phenomenon. Laclau and Mouffe define a discourse as the fixation of meaning within a particular domain: a "fixation of meaning around certain nodal points" (Phillips and Jørgensen 2002: 26). Lacan has called this a master signifier, a sign or a term that condenses the meaning of a series of other signifiers. Around a nodal point, there normally are a series of floating signifiers, loosely defined and in fact fixed through the way they are chained together and tied to nodal points (ibid: 27). This is exactly what happens to a term like "Arctic." By tying the term to remote, wild, primitive, and peripheral, the term is defined. There are several books that look at northern tourism from this angle. Hall in 2007, and together with Boyd in 2004, sums up the conception of tourism in peripheral areas: this is a tourism that tends to be far from the markets, lacks political control, has migration outflows, lacks innovation and state interventions, and has less information. Such statements may contend some kind of "truth," but are generalizations that at the best inaccurate? Rich on natural resources, reckoned as beautiful, and exploited are not applied terms. Thus, the interesting questions are why such signifying chains exist and who has the power to define it this way. In another book about tourism in peripheral areas, Hall and Boyd (2005: 6) insist on nature being essential for this activity but goes on writing: "Naturalness, sometimes also termed primitiveness . . . ." The chained signifiers loading the nodal point in the same part of the article are nature, primitiveness, wilderness, and remoteness. A term like "primitive" is not likely to be used by an insider. It is worth noting that nobody within the group of 15 tourism researchers from the area used the term "periphery." Also this is probably a result of a recognized competition of power to define, within the public, in politics, and in research. With reference to Laclau and Mouffe, Torfing claims that "every discourse is constituted as an attempt to dominate the field of discursivity by expanding signifying chains which partially fix the meaning of the floating signifiers" (1999: 98). The fixation here is a negative interpretation of "periphery" in the public.

However, within the sphere of tourism, an alternative discourse exists, for instance, in the tourism sector itself. Here, the nodal point is uniquely explained through the geophysical characteristics tied to touristic narratives related to exotic, beauty, untouched, and adventurous. This is the image the tourism industry tries to promote. But the public discourse is also part of the image. Public opinions are one of the strongest foundations for images (cf. Beerli and Martin 2004). Thus, concerning tourism, there are two competing. In the next section, the way this is handled within the tourism academia will be shown.

3.4 Sustaining Myths About the Periphery in Tourism Academia

Since Christaller (1964) in the 1960s stated that the peripheries were shaped for relaxation and recreation and Turner and Ash (1975) a decade later used the term the "pleasure periphery," several tourism books have contained "periphery" in their titles (Brown and Hall 2000; Hall and Boyd 2005; Müller and Jansson 2007). Etymologically, "periphery" means a location on the circle – and thus is defined in relation to the center. However, the term has its semiotic history that influences the ways it is interpreted. It is a "thick" concept or signifier (Huysmans 1998) filled with knowledge, myths, stereotypes, and prejudices. How the term is linked to other terms and phenomena is shown above. It is also cultural and contextual – and the meaning of periphery is not the same in the periphery as in the center. It has, not surprisingly, been claimed that those being occupied by the issue are those being the victims to the peripherization and that therefore the critique of the use of the term is based on feelings, more than facts (Rodríguez-Pose 2006, quoted by Best 2009).

In discussing the dealing with "periphery" in academia, Best (2009) presents three analytical models. The first and dominating is to relate the term to power: the center is the seat of power and the periphery is dominated by this center. This is the model applied also among most of those occupied by resisting this pattern. The solution of the "problems" of the periphery is empowerment (ibid). Gregson et al. (2003) applied this model in their analysis and critique of the academic hegemony within the field of geography. A second model relates to modernization: the center is the productive (and intellectual) core, those being the best are those that have taken the power positions; the Anglo-American scholars have hegemony because they are in the lead of academic work and their hegemonic positions are in fact deserved. In the third position, Best (2009) refers to those claiming that the center-periphery thinking is too simple as there are lots of academics from outside the Anglo-American sphere that have central positions in their fields. Thus, the academic space is not hierarchy or hegemony with a clear center, but a hybrid structure not with fixed borders, giving room for everybody that comes up with something valuable. This may be right, but still the majority of those publishing about tourism are from the Anglophone world (Dann and Parinello 2009), probably originating from the central areas of their countries.

The existing writing about tourism in the periphery is probably primarily of the second type, whereas the critique is of the first type – as in this chapter relating it to hegemony. But some of the writing can also be contested from a perspective of modernization, for instance, the article of Hall (2007) referred to above. Concerning these periphery characteristics listed by Hall, there are obvious variations and counteracting evidence, at least for the Arctic. The distance to the market is his first point. This is not necessary a problem; in Arctic tourism, it is in fact a selling point, and from a sustainability perspective, it reduces the tourist volume and reduces the pressure on the environment. Although there are huge areas in the Arctic that are rather inaccessible, lacking the infrastructure needed for tourism, other places are lot more accessible to Western (and particularly European) markets than Australia and New Zealand. A tourist host in the core of the Sami area puts this in perspective by saying that Maze, where he lives, in the middle of the Sami area, is in a central place, 3 h from Oslo, 10 h from New York. Concerning political control, there may be a lack of autonomy in many Arctic areas, but not necessarily less than in other regions. For instance, Svalbard in many ways is the most governed, monitored, and controlled area of Norway. However, the local community is strongly involved in this governance. Further, there is a net outbound migration in the Arctic, but many northern places like Alta, Tromsø, and Rovaniemi have growth. In fact there is an urbanization process going on in the Arctic, as in very many countries and regions on Earth, where the rural districts – often referred to as the periphery of the periphery – are the "loser." And there are lots of in-movers from the south – the periphery also has migration inflow. Hall also mentions lack of innovation. This obviously varies, but a book called *Innovations in the northern periphery* (Aarsæther and Bærenholdt 2004) was published a few years ago. The book was written and produced in the Arctic (Tromsø) and is about innovations in the Arctic areas of Scandinavia. State interventions as Hall also ties to the periphery occur all over, also in the periphery. But in many places, it is asked for. State compliance is not discussed by Hall, neither state-based nor center-based exploitation of the periphery. And his last point that the periphery is less informed is a doubtful stance. This is certainly not true for all people living there. Longyearbyen (with about 2,000 inhabitants) on Svalbard, for instance, is probably one of the towns on Earth with the highest average educational level. People there live off research, teaching, satellite operations, other services, and mining. Thus, all in all, such characteristics, referred to but not all acclaimed by Hall (2007), are producing an image of the periphery as a problem. And particularly this is the meta-message. On the other side, such presentations also show lack of knowledge, academic ignorance, or lack of methodological skill; such generalizations are mostly examples of aggregate faults and metonymic generalizations, at least seen from a Scandinavian point of view.

A contrast to such accounts is a book called Mobility and Place (Bærenholdt and Granås 2008), a book produced in the "less-informed periphery" with the subtitle Enacting Northern European Peripheries, emphasizing its "deviant" origin. The publisher is Anglo-American. The book's aim is to contribute "with more critical considerations of the notion of the periphery and the center-periphery dichotomy" (Bærenholdt and Granås 2008: 4). The book as such is probably easiest to place within the hybridity model (Best 2009). In one of the chapters, Paulgaard (2008) shows how young people reject the center-periphery dimension, ranking their companions according to scales of fashions known among young people everywhere. In other chapters, it is shown how Tromsø is linked to an international political scene (Kraft 2008) and how people on Svalbard adopt to a life of mobility (Viken 2008). In fact many of the chapters show that the life in the "periphery" is very modern and diverse. Most of "the periphery" is a mixture of remoteness and closeness, of the rural an urban, and of mobility and place. And is it necessary that the periphery is remote? An academic living in Tromsø already in the 1970s wrote a book she called "It is Oslo that is remote" (Skard 1974). The distances to the centers of power can be enormous also for people living in the centers. The point here is that space is hybrid (Woods 2005: 302), both in the center and the periphery. However, the periphery continues to be othered, particularly by those living in the center and also by tourism academics. Paulgaard (2008: 52) makes the following statement concerning the stamping of north as a periphery: "At a symbolic level, the hegemonic understanding of the northern periphery functions as an encounter with difference and the North becomes fundamentally different, a negation of civilized life in more central areas."

3.5 A Consequence of the Anglo-American Hegemony: A Biased Knowledge?

A series of problems occur when reading some of the Anglo-American literature about Arctic tourism. Here two elements will be discussed: inaccuracy and generalization. In a recent book called *Prospects for Polar Tourism*, Snyder and

Stonehouse (2007) present polar tourism. The volume comprises much valuable knowledge but is also an example of how it can turn out when people are not familiar with an area they write about. In one of the chapters, Snyder (2007b) looks at the economic significance of Arctic tourism. This presentation gives an odd impression. One of the problems is what he compares regions with countries. Concerning Canada, the focus is on the north; concerning the USA, Alaska is chosen; and for Russia, there is also a northern delimitation. In terms of Norway, Finland, and Iceland, the entire countries are included. Seen from the center, all Nordic countries are probably northern, but they are not Arctic. Less than 10% of Norwegians live in polar areas, and most of their southern compatriots have no knowledge of or identity related to these areas. In this, the author does not even follow the definitions of the Arctic given by himself and his coeditor in the beginning of the book. For Norway, Snyder (2007b: 112) gives some figures for development in which he probably is confusing millions and billions (can both be abbreviated as "mill." in Norwegian). He states that a little less than two million Euros (14 million NOK) were invested in tourism per year between 1997 and 2004. It is definitely wrong and clearly demonstrates distance and unfamiliarity with the area described. Concerning Arctic tourism attractions in Scandinavia, the book only mentions the Swedish and not the major successes that are to be found (see below). The impression this presentation gives is first and foremost that its author lacks knowledge of its empirical base, the Arctic. Another example of inaccuracy is found in articles written by Mason. He states that unlike Antarctica, there are indigenous people in the Arctic (Mason 2007). To him this makes a big difference. This is right, but he ignores ethnic Russians, Finns, and Norwegians. He also ignores the policy and regulation changes that have taken place, for instance, in Svalbard since the early 1990s, for instance, a new environmental act and an encompassing monitoring system have been introduced. Mason also uses references to himself as evidence, for instance, about the environmental impacts of tourism in the Arctic (Mason 2005: 182). At best, such references reflect lack of evidence. He also states that as tourism has expanded in the Arctic, a bastardized, inauthentic pseudo-culture has emerged (2005). He is of course free to mean so. But what does this mean? That the Arctic has been modernized? Would he prefer that these areas remained untouched by modernity, maybe as cultural "zoos" for tourists from the centers of the world? The problem is lack of respect for the areas written about, or lack of reflexivity, but also about editors, referees, and publishers are letting such accounts through.

Another problem is generalization. Talking about the Russian Arctic, Snyder (2007a) says that the area includes 44 distinct indigenous ethnic groups. Then he goes on writing about them as "they," treating them as one group. Similar generalizations are also done by Hall (2007: 25–26) who claims that "peripheries tend to" and goes on with the characteristics discussed in an earlier section. In another context, Hall and Boyd (2005: 10) talk about a "lack of success of tourism development in peripheral areas." There certainly are places in the peripheral Arctic without success in tourism, as in many central areas. But there are also many success stories, namely, Iceland, Svalbard, North Cape, Lofoten, Kiruna (Jukkasjärvi's ice hotel), and Finnish Lapland with its Santa Claus and ski tourism. Are these cases not included because their successes exclude them from being peripheral? Are successes national, and failures local and peripheral? And what is really successful tourist development? There is, for instance, a big difference between North Norway and North Finland in this respect. In North Norway, the major success story is about tourism as a minor industry that together with other industries, such as fisheries, agriculture, reindeer herding, mining, shipbuilding, oil and gas production, university training and research, and service production constitutes a society almost without unemployment. The point is that tourism is neither needed nor wanted as a major industrial activity in most places in North Norway. People from outside registering a modest growth will possibly interpret it as a failure. Finland, with a huge and heavily subsidized tourism, is by many seen as the success story.

The problems shown about some of the writing about Arctic tourism may also be an example of the importance of contextual knowledge; many of the faults would not have emerged, if the writers were properly informed about their empirical area. The problem relates to the fact that much writing about Arctic tourism seems to be based on knowledge about tourism in Antarctica. This is a peculiar point of departure. Whereas Antarctica has no (aboriginal) inhabitants, there are several million people living in the Arctic, both people identified as aboriginal and others, and the others constitute the majority. Furthermore, the Arctic contains modern and complex towns (with up to several hundred thousand inhabitants) and communities within which tourism takes place. Correctly then, unlike Antarctica, there some million people living in the Arctic, which makes challenges for the tourism industry quite different to those faced by the Antarctic and comparisons and knowledge transfer almost impossible. Probably also, therefore, the presentations of Arctic tourism referred to above are superficial, inaccurate, and misleading generalizations. There are lots of examples of metonymization, or metonymic generalizations: fact and myths about parts that are representing the whole. This is of course a challenge to avoid when one tries to write in general about a huge area or areas that almost merely have in common that they are located far from the centers of the world.

3.6 Concluding Remarks: Academic Reification of Northern Otherness

The inaccurate and othering writing about Arctic tourism is primarily a result of the general hegemony of Anglo-Americans within the tourism academic writing. People originating from Arctic areas do not write and publish much about their area; the tendency is that that these researchers tend to avoid the stigma of Arctic orientation and write in "general" terms, avoiding or hiding their northern empirical basis. Thus, there is an empty space concerning Arctic tourism. Some of those individuals, but not all, filling this space demonstrate arrogance in their writing by not really checking out the realities and without even citing any literature in their native language (cf. Aalbers and Rossi 2006). The majority of literature about

Arctic tourism indicates that those being familiar with Anglo-American publishers are defining the problems and challenges in the region. Most of these accounts are balanced and based on evidence. But as this chapter has shown, some are biased, adding to myths and stereotype images of the north.

What are the consequences of this type of othering? Is it not just good promotion? May be so, but there are some problematic sides. First, this is not how research should be; academic accounts should be as correct as possible. Although such accounts will not survive these days, the book probably occupies a scarce writing space for Arctic tourism. Research and academic writing constitute the knowledge base for students and future employees in the tourism sector. The university sector has a responsibility to provide these people with information that is as correct as possible – and avoid and break down myths. Future tourism industry should be based on proper knowledge. But because the knowledge provided – as is discussed here – often is dubious, the most important is to make future employees critical and reflexive, able to judge the value of the information with which they are confronted. Therefore, it is important that the discipline itself is self-reflexive.

Second, the writing practice referred to adds to reifying processes. Reifying means that prevailing opinions are taken for granted and looked upon as fact, correct, autonomous, impersonal, objective, and so on (cf. Thomason 1982:88). To refer to the Arctic as a periphery, pristine and primitive, is to make such a reification. The term also refers to culture conceived of as fixed structures that more or less give directions for people's lives (Keesing 1994). As shown, a strong public opinion exists about the north as less modern and civilized. Scholars claim this is wrong, that alternative modernities exist (Olsen 2008), and that the modern north is different from that of the south. And somehow, the hybridity that obviously exists is denied; hybrid space (Woods 2005; Best 2009) means a geography that rejects the relationship between a certain location and the type and quality of human activities enacted there. But this is neither accepted by the public nor by academics in general. Hall's (2007) article is an example. The northern tourism industry even takes part in these processes, often appreciating being regarded as the Other. There is an aestheticization and exotization going on in which tourism is a significant contributor that depoliticizes essentialization according to Kirschenblatt-Gimblett (1998: 76). The hegemonic processes are not easy to stop.

Are there any solutions? Aalbers and Rossi (2006) argue for establishing alternative publishing channels and see a growing post-national writing space in human geography, opposed to the Anglo-American one, as a step forward. In parallel, there is a growing northern academic tourism writing practice, producing authorized accounts from within, though there probably still is a lack of publication channels. Aalbers and Rossi (ibid: 145) also see signs of what they call "cosmopolitan project aiming at a decentered and situated universalism" and a more critical stand on the world's order in academia. This may be the most important way out of the problems discussed in this chapter, a more comprehensive reflexivity among researchers and writers within the Arctic, but also within Anglo-American academia and particularly among editors and publishers. Recent examples of coproduced research and publications within tourism academia may be taken as signs of movement in this direction. However, it is not only a question of reflexivity, it also has to do with morals and responsibility – well-known topics within academia – but obviously not easy to practice.

References

Aalbers MB, Rossi U (2006) Beyond the Anglo-American hegemony in human geography: a European perspective. GeoJournal 67:137–147

Aarsæther N, Bærenholdt JO (eds) (2004) Innovations in the Nordic periphery. R2004: 3. Nordregio, Stockholm

Bærenholdt JO, Granås B (2008) Mobility and place: enacting northern peripheries. Ashgate, London

Beerli A, Martin JD (2004) Factors influencing destination image. Ann Tour Res 31:657–681

Berg L, Kearns R (1998) America unlimited. Environ Plan D Soc Space 16:128–132

Best U (2009) The invented periphery: constructing Europe in debates about 'Anglo hegemony' in geography. Soc Geogr 4:83–91

Brown F, Hall D (eds) (2000) Tourism in peripheral areas. Channel View, Clevedon

Christaller W (1964) Some considerations of tourism location in Europe: the peripheral regions – underdeveloped countries – recreation areas. Papers in Regional Science 12(1):95–105.

Dann G, Parinello GL (2009) Setting the scene. In: Dann G, Parinello GL (eds) The sociology of tourism: European origins and development. Emerald, Bingley, pp 1–64

Flyvbjerg B (2006) Five misunderstandings about case-study research. Qual Inq 12(2):219–245

Foucault M (1980) Power/knowledge: selected interviews and other writings 1972–1977. Harvester Press, Brighton

Gibson C (2008) Locating geographies of tourism. Prog Hum Geogr 32:407–422

Gregson NA, Simonsen K, Vaiou D (2003) Writing (across) Europe: on writing spaces and writing practices. Eur Urban Reg Stud 10:5–22

Grenier A (2007) The diversity of polar tourism: some challenges facing the industry in Rovaniemi, Finland. Polar Geogr 30(1–2):55–72

Hall S (2001) Foucault: power, knowledge and discourse. In: Wetherell M, Taylor S, Yates S (eds) Discourse theory and practice: a reader. Sage, London

Hall CM (2007) North–south perspectives on tourism, regional development and peripheral areas. In: Müller DK, Jansson B (eds) Tourism in peripheries: perspectives from the Far North and South. CABI, Wallingford, pp 19–38

Hall CM, Boyd S (2005) Nature-based tourism peripheral areas: introduction. In: Hall CM, Boyd S (eds) Nature-based tourism in peripheral areas: development or disaster? ChannelView Publications, Clevedon, pp 3–17

Heller A (1999) A theory of modernity. Blackwell, Oxford

Huysmans J (1998) Security! What do you mean? From concept to thick signifier. Eur J Int Relat 4(2):226–255

Johnston M (1995) Patterns and issues in Arctic and sub-Arctic tourism. In: Hall CM, Johnston M (eds) Polar tourism: tourism in the Arctic and Antarctic regions. Wiley, Chichester

Keesing RM (1994) Theories of culture revisited. In: Borofsky R (ed) Assessing cultural anthropology. MacGraw-Hill, New York, pp 301–312

Kirschenblatt-Gimblett B (1998) Destination culture: tourism, museums and heritage. University of California Press, Berkeley

Kitchen R (2005) Disrupting and destabilizing Anglo-American and English language hegemony in geography. Soc Cult Geogr 6(1):1–12

Kraft SE (2008) Place-making through mega-events. In: Bærenholdt JO, Granås B (eds) Mobility and place: enacting northern peripheries. Ashgate, Aldershot, pp 219–232

Maher P (2007) Arctic tourism: a complex system of visitors, communities, and environments. Polar Geogr 30(1–2):1–6
Mason P (2005) Visitor management in protected areas of the periphery: polar perspectives. Tour Hosp Plan Dev 2(3):171–190
Mason P (2007) Visitor management in protected areas of the periphery: experiences from both ends of the world. In: Müller DK, Jansson B (eds) Tourism in peripheries: perspectives from the far North and South. Cabi, Wallingford, pp 154–174
Meriläinen S, Tienari J, Thomas R, Davies A (2008) Hegemonic academic practices: experiences of publishing from the periphery. Organization 15:584–597
Müller DK (2009) Nordic tourism: is there a Nordic way? 18th Nordic symposium in tourism and hospitality research, University of Southern Denmark, Esbjerg, October 22–24 2009
Müller DK, Jansson B (eds) (2007) Tourism in peripheries: perspectives from the far North and South. Cabi, Wallingford
Olsen K (2008) Identities, ethnicities and border zones: examples from Finnmark, Northern Norway. Ph.D. thesis, University of Bergen, Bergen
Paasi A (2003) Region and place: regional identity in question. Prog Hum Geogr 27:475–485
Paulgaard G (2008) Re-centring periphery: negotiating identities in time and space. In: Bærenholdt JO, Granås B (eds) Mobility and place: enacting northern peripheries. Ashgate, Aldershot, pp 49–59
Phillips L, Jørgensen MV (2002) Discourse analysis as theory and method. Sage, London
Rodríguez-Pose A (2006) Is there an'Anglo-American' domination in human geography? And, is it bad? Environ Plan A 38:603–610
Skard T (1974) Det er Oslo som ligger avsides: glimt fra en reise i norsk utkantpolitikk – Troms og Finnmark. Pax, Oslo
Snyder JM (2007) Managing polar tourism: issues and approaches. In: Snyder JM, Stonehouse B (eds) Prospects for polar tourism. Wallingford, Cabi
Snyder JM (2007a) Managing polar tourism: issues and approaches. In: Snyder JM, Stonehouse B (eds) Prospects for polar tourism. Cabi, Wallingford, pp 231–246
Snyder JM (2007b) The economic role of Arctic tourism. In: Snyder JM, Stonehouse B (eds) Prospects for polar tourism. Cabi, Wallingford, pp 102–121
Snyder JM, Stonehouse B (eds) (2007) Prospects for polar tourism. Cabi, Wallingford
Stewart EJ, Draper MD, Johnston ME (2005) A review of tourism research in the polar regions. Arctic 58(4):383–394
Thomason BC (1982) Making sense of reification. Macmillan, London/Basingstoke
Torfing J (1999) New theories of discourse: Laclau, Mouffe, Zizek. Blackwell, Oxford
Turner L, Ash J (1975) The golden hords: International tourism and the pleasure periphery. Constable, London
Urbinati N (1998) From the periphery of modernity: Antonio Gramsci's theory of subordination and hegemony. Pol Theory 26:370–391
Viken A (2008) The Svalbard transit scene. In: Bærenholdt JO, Granås B (eds) Mobility and place: enacting northern peripheries. Ashgate, Aldershot, pp 139–154
Woods M (2005) Rural geography. Sage, London

Chapter 4
Narrative Frameworks of Consideration: Horizontal and Vertical Approaches to Conceptualizing the Sub-Antarctic

Eric J. Shelton

Abstract Tourists engage with narrative as they experience location, construct place, and act or perform within produced space. Northern and southern polar and subpolar regions offer multiple opportunities for narrative engagement. When conceptualizing tourists' narrative engagement with location, place, and space in the sub-Antarctic and in Antarctica, it is interesting to consider tensions, consistencies and inconsistencies, and coherence and incoherence produced when activities and processes occurring on this part of the earth's surface are grouped and analyzed either vertically or horizontally. Many engagements with the area are organized vertically, along two axes. One vertical axis is through Southern Australia, New Zealand, and various islands to the Ross Sea. Another vertical axis runs from the tip of South America, through various islands, to the Antarctic Peninsula. Alternatively, the sub-Antarctic may be conceived of horizontally as a distinct circumpolar domain lying between specified latitudes. Both of these analytic approaches, vertical and horizontal, involve the sub-Antarctic being described in ways which are themselves narratives of appropriation in that the descriptions are framed as persuasive communications intended to privilege one approach over the other.

Academic study of sub-Antarctic tourism traditionally has embraced a horizontal approach, while the industry is organized vertically. This chapter examines the implications of this divergence for tourists' personal production of narrative, the provision of public and normative narratives, and the development of a compelling sub-Antarctic tourism research agenda.

Keywords Sub-Antarctic • Worldmaking • Narrative • Self • Subjectivity

---

E.J. Shelton (✉)
Department of Tourism, School of Business, University of Otago, Dunedin, New Zealand
e-mail: eric.shelton@otago.ac.nz

D.K. Müller et al. (eds.), *New Issues in Polar Tourism: Communities, Environments, Politics*, DOI 10.1007/978-94-007-5884-1_4,

> [Worldmaking] is the creative and often 'false' or 'faux' imaginative processes and projective promotional activities that management agencies, other mediating bodies, and individuals strategically and ordinarily engage in to purposefully (or otherwise unconsciously) privilege particular dominant/favoured representations of people/places/pasts within a given or assumed region, area, or 'world', over and above other actual or potential representations of those subjects. (Hollinshead 2009, p. 140)

In this chapter, I will take Hollinshead's notion of the worldmaking potential of tourism and situate it within a current debate about the relationship between Antarctica and the sub-Antarctic. This debate pertains to all aspects of this relationship, not just tourism, but here I will consider only tourism and I will demonstrate that the current vertical sub-Antarctic spatial organization of the industry is unlikely to change. By considering wider issues of self, subject, and narrative, it is possible to suggest that if worldmaking is to take place, still it will be organized vertically and based only on the sub-Antarctic islands for which Australia and New Zealand are responsible, not those islands presently associated with travel to the Antarctic Peninsula. I illustrate how tourist subjectivity, as a tension between "the subject who knows and who speaks" and "the subject who is an effect of power, science or technologies" (Fensham 2000, p. vi), operates within the imaginaries that such worldmaking produces.

4.1 The Current Position: "Sub-" as in Subaltern

Postcolonial writing supplies an apt descriptor for the current relationship between the sub-Antarctic and Antarctica, subaltern. Through the vagaries of its history, exploration, settlement, and past use (Russ 2007), the sub-Antarctic region has come to be classified through what it is purported to be lesser to, Antarctica. Certainly, Antarctica determines the region's climate and weather, which in turn determines many features of the sub-Antarctic's natural history and current use. Miller (2011) argues that this "relational descriptor," sub-, is an impediment to any attempt to have the region classified as a cohesive whole and should be dispensed with[1] Boyer and Haywood (2006), for example, use the term "islands of the Southern Ocean" to describe occupancy of the four horizontal biogeographical zones in contention. The Tasmanian *Department of Economics, Tourism and the Arts* (2011) has adopted the term "Antarctica and the Southern Ocean," which offers a context within which to situate the islands. Part of the drive to have the islands recognized as a circumpolar region involves "(f)inding a new name for the region consonant with its status as a distinct entity rather than an appendage of Antarctica ... perhaps the time has come to sever the Antarctic association entirely and think in terms of something like *The Great Southern Zone*" (Green 2005, p. 181).

---

[1] The most comprehensive list of sub-Antarctic islands is to be found at http://en.wikipedia.org/wiki/List_of_Antarctic_and_subantarctic_islands and a useful map is at http://www.sub-antarctic.org/docs/Subantarcticmap.pdf

What would inform the choice of a replacement term is open to debate and raises the question of what feature or features define the region. Walton and Dingwall (1995) offer one description: "Outside the Antarctic Treaty area but well within the influence of the Antarctic continent lies a ring of remote islands. These subantarctic islands . . . " (p. ix). Significantly, "(t)here is no real consensus as to what constitutes the sub-Antarctic but there is a great deal of opinion in common" (Selkirk et al. 2007). No definitive characteristic can be applied to the land masses of the region; the most widely shared feature is that they are all true islands, that is, they are not and never have been part of a continent. South Georgia is an exception. The islands are predominantly volcanic, with three exceptions, and they are remote, with one exception. This tiny exception, Solander Island, is only 200,000 years old and is situated just 40 km off the southern New Zealand coast (Lewis and Landreth 2010). Although of interest to scientists, Solander is ignored by commercial tourism operators and is not mentioned in Peat's (2003) guidebook on New Zealand's sub-Antarctic islands. The island, more accurately two small islands, offers neither remoteness nor spectacle but being so recent is of geological interest. Mels (2008) notes that "environmental discourses draw attention to how the production, circulation and justification of meaning within particular constellations of power permeate all social practices and thereby always enter into the constitution of the biogeophysical environment" (p. 387). This process is illustrated by the descriptors commonly in use about the sub-Antarctic.

So, "island-ness" (Baldacchino 2006; Halpenny 2001) and biogeography offer support for the notion of the sub-Antarctic as a circumpolar band. Gibbs (2006) points out that the fauna of the region, represented by penguins, are there since:

> polar bears are in the north because that is where bears evolved and penguins claim the south because that is where they evolved . . . Neither group has dispersed into the other's territory . . . even though they would probably be perfectly capable of living there. (Gibbs 2006, p. 40)

The factor that most strongly militates against the establishment of a circumpolar region is governance. Each of the sub-Antarctic islands is governed by a noncontiguous sovereign state, and there is no supranational organization that can assume governance functions, despite there being numerous "international instruments and arrangements" (Jackson 2006) available and in use. None of these instruments and arrangements has the power of the Antarctic Treaty, where certain aspects of sovereignty are ceded for the greater good of the continent. With respect to governance activity by NGOs, Landau (2007) discusses how the International Association of Antarctic Tour Operators (IAATO) has expanded its coverage to sub-Antarctic tourist activity "north of 60°S to the boundary of the Antarctic Convergence, or Polar Front, in order to include . . . South Georgia, Îles Kerguelen and Heard Island" (p. 174). Still, this expansion of operations retains Antarctica as the focus and central to policy formation and the adoption of concepts. For example, ambassadorship (Maher et al. 2001), an idea originating in Antarctica, was suggested by Liggett (2011), while presenting the sub-Antarctic as an economic and environmental asset, as something we have learned in Antarctica that may usefully be applied in the sub-Antarctic environment. These contributions, both by Landau and Liggett, remain focused primarily on Antarctica and treat the sub-Antarctic as an extension of that continent. This difficulty in changing focus demonstrates the size of the conceptual leap that is required in order fully to overcome Antarctica's influence.

The *International Forum on the Sub-Antarctic* is a venue that facilitates advocacy and challenges the privileging of Antarctica. Scientific meetings have been held in Hobart, Tasmania, in 2006, 2009, and 2011.

> The forum was founded upon the conviction that the Sub-Antarctic islands and surrounding seas should not be viewed as a subdivision of Antarctica but as a distinct circumpolar domain which has its own unique character, scientific significance and management issues. (Antarctic Tasmania 2011)

It is difficult to establish exactly why the reconceptualization of the sub-Antarctic is an idea whose time may have come. Individual scientists who attend the forum clearly see benefits to science if research is funded and conducted within horizontal rather than vertical parameters. In contrast, the sub-Arctic region faces pressing social as well as ecological issues, stemming directly and indirectly from climate change and requiring human-focused interventions (Hull 2011). These northern hemisphere social issues revolve around the rights of indigenous peoples and are absent in the south. In the south, the largest movement of people is due to tourism. What would need to happen for tourism to play a role in developing a circumpolar domain? How would such a development occur, and how likely is such a change of emphasis? In order adequately to address these questions, it is essential to interrogate the idea of the tourist as subject, situated within the appropriate tourist system.

4.2 The Current Sub-Antarctic Tourist System

Currently, the tourism system involving the sub-Antarctic is organized vertically (Shelton 2007, 2011). Ships travel south from Ushuaia in Argentina to the Antarctic Peninsula and visit sub-Antarctic islands along their way (Bertram et al. 2007). A few ships forego the Antarctic Peninsula and visit only the larger sub-Antarctic islands. The other vertical axis is from Hobart, Tasmania, and Dunedin or Bluff in southern New Zealand to Ross Sea, taking in the Snares, Campbell Island, the Auckland Island group, and Macquarie Island. This vertical arrangement is not just a spatial organization of the region but also is conceptual. For those who advocate for a horizontal circumpolar focus, the issue is how to move from vertical to horizontal, conceptually and practically. There are existing examples of the implications of particular spatial formations. Hall et al. (2009) proposed that there is a permeable boundary between "Nordic tourism space" and "non-Nordic tourism space" and that this boundary involves a recognition of the spatial formation of identity. In a similar way, the spatial arrangement of the sub-Antarctic has shaped personal and group narratives of place, space, self, and identity among tourists (Shelton 2007).

4.3 The Sub-Antarctic Tourist: Self and Subjectivity

What is this notion of identity that Hall et al. (2009) and Shelton (2007) refer to? Before attempting to answer that question, it is necessary first to engage with a notion constitutive of and constituted by identity – that of selfhood.

> 'The self' stands for a very particular understanding of personal identity, one that presupposes an essential core of selfhood characterized by psychological depth, or interiority, which is the bedrock of unique, expressive individual identity. (Wahrman 2004, p. xi)

Although our interest in self and identity is limited to tourism in late modernity, the topic has been the focus of long-standing enquiry.

> (T)he nature and meaning of selfhood have been recurring questions . . . in practically every known human time and place. Nowhere has the debate been more full-blown or more intense than in the modern West, the locale in which individuality has been most fervently celebrated and most ardently denounced. (Seigel 2005, p. 1)

Such individuality informs the current neoliberalism of the West. The tourism industry is well suited to such a political and economic environment. One of the features of neoliberalism is the putative freedom to govern one's life by making personal decisions. One such decision is how to narrate oneself.

> Of all the constructs the modern era championed, none has been more problematic for people in the postmodern era to consider than 'the self' . . . Kenneth Gergen (1991) in *The Saturated Self* asked us to reflect upon our 'romanticized notions' of a coherent and ever agentive self . . . narrative understandings of self opened up a compelling view of lives lived in a flow of constructed meaning . . . lives that could be appreciated in historically intelligible ways . . . though analyzing lives in narrative ways always posed questions about the comprehensiveness of any story of the self, and the contextuality of such stories. (Strong 2000: np)

Part of being able to narrate oneself involves being able to recall past episodes involving oneself:

> If our identities derived solely from where we were at any one time, we would become extremely confused, not to say mentally exhausted with the effort of placing ourselves. Thus we construct for ourselves workable biographies, reflections upon our lives through which we interpret our aging as it happens. Each of us needs to connect with our own past. At the same time we can only assemble our agency from the resources locally available to us. (Blaikie 2005, p. 167)

Repeatedly constructing workable biographies and assembling agency work to produce a sense of identity:

> Identity is the sense of who one is in the world, distilled from a lifetime of experience. From the collective perspective, identity consists of the life course as a cultural construct: socially normative and collectively outlined and accepted life-course statuses and transitions. But at the individual level, every person creates for herself a particularised version of the collective life course, a life story, depending upon her specific experiences and the meanings she attaches to them. (Rubinstein 1992, p. 144)

This process is the embodiment of subjectivity, and this embodiment contains the germ of the tourist as consumer. Morton (2007) suggests when this role became available:

> By the Romantic period, it became possible to be (or if you prefer, to *act*) consumerist. Consumerism is a reflexive mode of consumption. It is about how one appears as a certain type of consumer... Consumerism raised to the highest power is a free-floating identity, or identity in process. This is a specifically Romantic consumerism. Transformative experiences are valued, such as those derived from drugs, or from intense experiences ... traumas that nudge the self out of its circularity and force it to circulate around something new. (Morton 2007, p. 111, italics in the original)

For many, visiting the sub-Antarctic is such a transformative, intense experience. In his *Land of the Blue Sunflower*, Fell (2002) provides an individual reminiscence:

> Today, the Galapagos Islands are no longer the lost world that Darwin discovered ... Everywhere on Earth, it seems, the spread of humankind has shrunk or tainted the wilderness... It became an obsession of mine to see a blue sunflower growing on the slopes of one of the most isolated islands on earth" (Fell 2002, p. 13) ...like a meadow garden orchestrated by some superhuman power" (p. 58).... it is the vast alpine meadows spangled with megaherbs and tussock grasses, that keep calling me back to Campbell Island. Not just the flowers themselves, impressive though they are, but a powerful feeling of being surrounded by turbulent ocean and being close to the dawn of creation (p.132)... Also special is the sublime silence, and the purity of the atmosphere constantly cleansed by wind and rain. All these sensations induce a strong, spiritual, uplifting state of peace and contentment, a strong sense of freedom.... (p. 133)

Fell (2002) thus offers the blue sunflower wrapped in purple prose and allows his imagination to wander, supposing that Disney World has leased the island; "Dear God...please keep Campbell Island as unspoiled as possible" (p. 63). He is not alone in being enamored of the island: "Glad I am that I came, I go home a different person" (Birchall 1996, cited in Fell 2002, p. 134). Although Fell's is an individual example of Romantic consumerism briefly being satisfied by visiting Campbell Island:

> (s)ubjectivity is not simply an individual, and certainly not an individualist, phenomenon. It is a collective one. (*sub-Antarctic cruise tourism*) is a way of registering the feeling of being surrounded by others (*other nature tourists*), or more abstractly, by an otherness (*a sublime landscape and charismatic flora and fauna*), something that is not the self (*but the discursive character of the narrated self comes close*). Although it may displace the actual social collective and choose to (*look and photograph and talk about*) surrounding mountains instead, such displacements always say something about the kinds of collective life that (*the green, environmentalist, deep ecology, Gaia, biocentric, ecofeminist, neoliberal, tourist*) is envisioning. (Modified from Morton 2007, p. 17, italicized inserts mine)

Jamal and Hill (2002) point out "(w)hen it comes to the tourist's experience, the search for authenticity is an intersubjective, emotional, ethical/moral and spiritual quest" (p. 101). Such is Romantic consumerism, and Antarctica and the sub-Antarctic islands both are well placed to provide the sought-after experiences. My position is that such performances, of nature and of tourism, may be generated by already available narratives, both of environment and of self. Such narratives may be produced outside of the individual person, in the form of a dominant public narrative

(Somers 1994) or as a normative narrative (Rappaport 1993). These externally generated narratives produce spaces of performance and make available various subject positions to be performed within these spaces.

> A subject position incorporates both a conceptual repertoire and a location for persons within the structure of rights for those that use that repertoire. Once having taken up a particular position as one's own, a person inevitably sees the world from the vantage point of that position and in terms of the particular images, metaphors, storylines and concepts which are made relevant within the particular discursive practice in which they are positioned. At least a possibility of notional choice is inevitably involved because there are many and contradictory discursive practices that each person could engage in." (Davies and Harre 1999, p. 46).

How then could sub-Antarctic tourism be made circumpolar? One way would involve making the subject position that hailed the Romantic consumer, in this case the tourist, uniquely of the sub-Antarctic, without reference to Antarctica. Given that the islands of the circumpolar band proposed by the *Forum on the Sub-Antarctic* have no single material feature in common, it is acceptable and necessary to treat the circumpolar concept as imaginary.

4.4 A Narrative Framework: Two Imaginaries

Shelton (2007) described the role of narrative in visitors' experiences of Heritage Expeditions' *Forgotten Islands of the South Pacific*, an expedition cruise tourism product that visits the sub-Antarctic islands south of New Zealand. Somers's (1994) concept of the *public narrative* proved useful in describing how a unitary narrative of environmental restoration was delivered, through the complimentary provision of Peat's (2003) book *Subantarctic New Zealand: A Rare Heritage* to every passenger, coupled with complementary onboard lectures and commentary. Passengers were left in no doubt about how they could most profitably engage with their surroundings. Nevertheless, some passengers had other reasons for being on the voyage and articulated these both through their "*ontological narratives*... the stories that social actors use to make sense of- indeed, to act in- there lives" (Somers 1994, p. 618) and how they deported themselves while onboard and while ashore. *Solander* is a suitable name for the imaginary in that it can accommodate multiple interpretations of the materiality of the region. *Solander* is apt since it is the name given to the area by the maritime weather forecast agency, NZMetService. *Solander* involves "(t)he semiotics of tourist spaces, landscapes and destinations" and "(t)he discursive construction and representation of the tourist experience" (Jaworski and Pritchard 2005) informed by a normative narrative of the Romantic and an aesthetics of the sublime.

Within narrative epistemology, there are no truth claims existing outside of the language used to produce them. "(T)he subject (Romantic consumer as sub-Antarctic tourist) is formed within language" (Ashcroft 2001, p. 38), so situating the tourists' *ontological narratives* and thus Law's (2000) "performative character of storytelling and its material embodiments," that is, stories of being a tourist bringing about being a tourist, required for this production of an imaginary, this act of *worldmaking*. When the nature tourist subject engages in Hollinshead's *worldmaking*, it is open to debate whether an essentialist, immanent Kantian sublime world is made or a sublime world more akin to one filled with Burke's superlative objects.

> Such is the declarative and clearly pungently political force of tourism as it is deployed in worldmaking fashion in concert with (or at times, wholly against) other co-productive and co-generative narrative-issuing mediating forces in and across society. (Hollinshead 2009, p. 140)

At the scale of the individual, the Kantian nature tourist and the Burkean nature tourist are such narrative-issuing mediating forces. At the scale of the firm, also the tourism operator is a narrative-issuing force, as are higher-level organizations. Narrative-issuing at all of these levels is implicated in producing the touristic subject. Ontological narratives such as those above may be sparked by a childhood reading of a particular book. For older New Zealanders, *Castaways of Disappointment Island* (Escott-Inman and Prater 1911) has this power to spark a consequent desire not only to visit that island but also to engage affectively and intellectually with the wider imaginary that reading this book, and subsequent books, engendered. This ontological narrative of desire to visit must be able to be constitutive of valid, more general knowledge claims. These knowledge claims involve, in this case, establishing how individual passengers' desires for the imaginary, constitute, and are constituted by other passengers' similar desires and how their desire is performed and narrated. The process for this generating of these knowledge claims, within language, is known formally as "interactive narrative development" (Quasthoff 1997, p. 51), involving "conversational narrative" (Quasthoff 1997, p. 58). It is usual that:

> (f)ollowing terminological conventions in linguistics, narrative discourse units are distinguished from other kinds of discourse units (e.g. argumentative, explanatory, descriptive) by both content and form. In terms of content, they refer to a singular event (as opposed to a recurring procedure) that happened in the past in which the narrator was involved in the role of either agent or observer. In terms of form, there are (at least) two possible global forms (or discourse patterns) in which both narrator and listener, in a joint achievement, realize the narrative discourse unit on the linguistic surface: the report pattern and the replaying pattern. The latter actualizes the past event strictly from the perceptual and experiential perspective of the narrator in the participant's role. (Quasthoff 1997, pp. 58–59)

Clearly, first-person accounts like Fell's (2002) about Campbell Island, retold in tour company brochures and websites, serve to produce such narrative discourse units. These narrative units offer ways of constructing new ontological opportunities.

> These worlds can be different than our own. Importantly, this means that story offers a new way of being in a world by providing the opportunity to envision environments and experiences through another's eyes or ears." (Hatton et al. 2010, p. 202)

When constructing an imaginary such as *Solander*, it is important not to be limited by geographical considerations of location. "These sub-Antarctic islands, although interesting in themselves, are *en route* to Ross Sea" reflects the narrative construction used by Australian providers in their marketing (e.g., Aurora Expeditions 2011) and fits better within a *Mawson* imaginary, at least around those products associated with the centenary of his expedition, specifically *Mawson's Centenary Celebration*. Any figurative relationship between *Solander* and *Mawson* of course must be constructed by the individual tourist, the embodied Romantic consumer. This lack of adherence to location makes clear that these imaginaries are systems of meaning and thus open to critique.

> I do, however, distinguish between postmodernism, as a cultural and ideological form, and *deconstruction*... which searches out, with ruthless and brilliant intensity, points of contradiction and deep hesitation in systems of meaning. (Morton 2007, p. 6, italics in original)

This idea of the deconstruction of personal identity narratives may be applied to wider issues of adhering to a social category because of one descriptive feature. Somers (1994), for example, challenged the idea that:

> (a)ll the members of a single category of actors... should behave similarly and have the same interests with respect to [in her example citizenship; in ours sub-Antarctic cruising] regardless of differences of residence, family or gender... But why do we premise or limit our understanding of people to their work category? Why should we assume that an individual or collectivity has a particular set of interests simply because one aspect of their identity fits into one social category- in this case their place in the production process? To let "class" stand as a proxy for experience is to presume what has not been empirically demonstrated- namely that identities are foundationally constituted by their categorization in the division of labor. (p. 624)

4.5 Narrating the Tourist as Embodied Subject

For our purposes here, "members of a single category of actors" refers to expedition cruise nature tourists in the sub-Antarctic. But does such a collection of individuals, considered *en masse*, where expedition cruising is only one aspect of each of their identities, constitute a social category? McCabe (2009) argued that:

> the idea of a *tourist* has taken on cross-cultural and cross-contextual ideological significance as a pejorative term with implicit political and moral implications in its use... *Tourist* as a categorization device can be subjected to analysis.... (p. 40, italics in the original).

Such a categorization device produces expedition cruise nature tourist as an identity that, although easily enough imposed on the traveling individual, may or may not be accepted by her, within the parameters of the cruise, or afterward. We may modify Somers' statement above to read, to let "sub-Antarctic expedition cruise nature tourist" stand as a proxy for experience is to presume what has not been empirically demonstrated, namely, that identities are foundationally constituted by their categorization in the division of labor. Somers (1994) then offered a solution by suggesting that:

> (s)ubstituting the concept of narrative identity for that of interest circumvents this problem. A narrative identity approach assumes that social action can only be intelligible if we recognize that people are guided to act by the structural and cultural relationships in which they are embedded and by the stories through which they constitute their identities ... people act in particular ways because not to do so would fundamentally violate their sense of being at that particular time and place. In another time or place, or in the context of a different set of prevailing narratives, that sense of being could be entirely different because narrative identities are constituted and reconstituted in time and over time. (p. 624)

For Somers, individuals are able to be their own biographical authors, the neoliberal position, but only when the structural and cultural relationships within which they are embedded already allow such authorship or such authorship is engendered by an act of resistance, and the narrative identity that results is ontologically acceptable to the individual, in a specific circumstance. Somers presented the notion of relational setting where:

> a relational setting is a pattern of relationships among institutions, public narratives and social practices. As such it is a relational matrix, a social network. Identity-formation takes shape within these relational settings of contested but patterned relations among narratives, people and institutions. (Somers 1994, p. 626)

The expedition cruise ship in the sub-Antarctic provides such a relational setting, metonymically related to wider tourism and in turn social and economic structures. The ship is the institution. I include in this term, institution, both the physical structure of the ship, the procedures used to operate it, the economic and legislative setting in which it operates, and the social practices that occur within it. The patterned relationships are between all of the individuals on board, and the narratives are a collection of tourists' ontological narratives, their narratives of self and identity, the subject positions operating and available to be filled, and the public narrative provided by the operator, and by the state, through regulation. The expedition cruise ship constitutes one kind of institution or structure. Another institution or structure is class. In Somers' sense, class can be an institution, being a determinant of attitude. "Different class structures influence the people caught up in them to form different attitudes toward nature and one another, and these attitudes likewise affect the natural environment" (Vlachou 2001, p. 105). Class, though, may be represented by niche market. "(V)isitors to these islands... were also characterized by higher proportions of older people, retired and professional people, women, and people involved in conservation groups" (Cessford and Dingwall 1998, p. 101).

To the extent that attitudes, above, are articulated through personal identity narratives, this position places any story of an imaginary, for example, the reasons for its extraordinary flora and fauna, firmly within discourse and such a combination, of story and politics, constitutes narrative. I am not claiming here that attitudes are closely related to performance or behavior more generally; what I am claiming is a strong relationship between attitudes and personal narrative, since attitudes must in some way be articulated, and it is this articulation, being of necessity political, that constitutes narrative. Lanchester (2011) commented:

> (b)ack when I was at university the only people who ever used the word 'narrative' were literature students with an interest in critical theory. Everyone else made do with 'story' and 'plot'. Since then the n-word has been on a long journey towards the spotlight- especially the political spotlight. Everybody in politics now seems to talk about narratives all the time... We no longer have debates, we have conflicting narratives. (p. 24)

Lanchester may be perspicacious in being early to spot a concept currently being devalued but "(m)ost narratologists agree that narrative consists of material signs, the discourse, which conveys a certain meaning (or content), the story, and fulfill a certain social function" (Ryan 2007, p. 24).

4.6 Worldmaking and the Imaginary

> [worldmaking] is the creative and often 'false' or 'faux' imaginative processes and projective promotional activities that management agencies, other mediating bodies, and individuals strategically and ordinarily engage in to purposefully (or otherwise unconsciously) privilege particular dominant/favoured representations of people/places/pasts within a given or assumed region, area, or 'world', over and above other actual or potential representations of those subjects. (Hollinshead 2009, p. 140)

It was clear that for the passengers engaged with *Solander*, the sub-Antarctic islands in question, the Snares, Campbell Island, the Auckland Island group, and Macquarie Island, offer an opportunity for placemaking, when these passengers, by their knowledge of the discourse, possess already a sense of place. This placemaking inevitably involves some form of performance of self or subject. The literature on tourist performance (Crouch et al. 2001; Crouch and Desforges 2003; Edensor 2001; Perkins and Thorns 2001) is marked by a move from a focus on gazing to focusing on other types of involvement. Sensory and perceptual involvement has been expanded to include smell (Dann and Jacobsen 2003), while cognitive and affective involvement has expanded to involve such imaginaries as *Solander*, that is, ideas and desires associated with, but not defined by, geographical and political entities and transforming them into touristic places and spaces:

> (W)hat places, and formative processes of place, generate and sustain significant desire, what are their material landscape qualities and how should we theorize and narrate their conditions? ... these places, their landscapes, and even their histories, are dynamic and contested, changing in relation to transformations in society and economy... whose draw owes to multiple sites of possible experience and sensory encounter.(Cartier 2005, p. 2)

Central to Cartier's considerations of desire formation is the notion of the

> touristed landscape... places whose larger, a priori significance arguably initiates desire to experience, tour, travel, and explore, rather than those where tourism economies have been explicitly created... This landscape is a place where locals and visitors negotiate identity, seeking renewal or exploration, the possibilities of alterity and liminal identity

> shift. Understanding experiences and meanings of the touristed landscape depends in part, then, on understanding subject positions and subject formation of the touring, the toured, and those who would work at being both or neither, and from one moment or place to the next. In such contexts, ideas people hold about places substantially inform identity formation, human agency, and questions of subjectivity. (Cartier 2005, pp. 3–4).

Here, the *tourist* fits within McCabe's (2009) notion of a categorization device, the embodiment of a subject position. This discursively produced tourist is, as above, "the subject who knows and who speaks" and is also "the subject who is an effect of power, science or technologies" (Fensham 2000, p. vi). In *Solander*, desire is central to considerations of narrative engagement with the sublime. Also, the sensory, perceptual, affective, and cognitive components of narrative are available to be appropriated by economic ideologies and then individuals interpellated into the resultant subject positions. Interpellated, as used here, is consistent with the original meaning of figuratively being "hailed," in this case being hailed to be a tourist and to perform tourism. This process legitimately is able to be subsumed within Hollinshead's notion of *worldmaking*. The fact that tourism at the beginning of the twenty-first century is the archetypal industry governed by neoliberal principles means that figurative narratives generated around that industry, whether by tourists, providers of tourism products, environmental managers, or organizations providing governance, will to a certain extent fit within a neoliberal worldview or will self-consciously offer resistance to it.

Stewart (1984) introduced the related concept of *longing*, which Bendix (2002) later phrased as the question:

> How can we explain ... craving for the new? Longing is encoded in objects that contain within them narratives or memories of experiences had or wished for ... Within longing, though, there are aspects that are inexpressible, either in narrative or objects ... Naturalists and explorers as well as philosophers of the sublime in the 18th century assisted their contemporaries in uncovering the affective possibilities of sights hitherto unseen. (Bendix 2002, p. 472)

These objects in which the longing is encoded clearly include the human body. Narratives, as well as being spoken and written, are able also to be presented within some other artistic medium, including bodily performance. "Human bodies do not just experience stories in space, they also interact within the materiality of narrative texts" (Page 2010, p. 10). Page's claim opens up the issue of the status of any truth claims implicit in such bodily performance; is it a performance of self or subject? The neoliberal metanarrative of *choice* clearly establishes an opportunity for the production and performance or production through performance of narratives that ultimately are presented as the product of the tourists' choices of how possibly to represent the contemporary sub-Antarctic as a set of tourism products differentiated from Antarctica. Such selection of narratives is another form of *worldmaking* where the choices clearly are informed by politics; Chaudhuri (1995) uses the term "Geopathology: The painful politics of place" (p. 55) to introduce the notion that:

> The problem of place- and place as problem- informs realist drama deeply, appearing as a series of ruptures and displacements ... the most fundamental dislocation is ... that between humankind and nature, which ... makes nature a mere setting- "scenery." (Chaudhuri 1995, p. 55)

The performance of the narrative self and that of the tourist subject within *Solander* is such realist drama; tourists' bodies jointly and severally move in location, place, and space according to sets of rules. What is the best figure for such performances?

> That figure is the theater, by which I mean both the theater in its usual sense as a dramatic art of the stage as well as its figurative sense as a stage for the performance of various modes of subjectivity, various calls or interpellations that transform the spectator into an actor... Performativity, then, is a tropic-concept that attempts to represent the process of presentation itself... Such a trope... offers a way of extending the seemingly innocent questions proper to philosophical discourse into certain political questions. The way we conceive of the social or political space in which we stage the ideological performances proper to certain political questions will influence the ways we conceive of those political questions themselves. (Hartley 2003, p. 182)

Hartley's comment is crucial for the proponents of a circumpolar approach to conceptualizing the sub-Antarctic. The social and political space the *Forum on the Sub-Antarctic* has created influences the way the political questions about the role of Antarctica are conceived. When it comes to producing a separate sub-Antarctic aesthetic, Morton (2007) is in no doubt that politics is involved:

> A consideration of the aesthetic is vital, since the aesthetic intertwines with the idea of the surrounding environment or world. The idea of a 'good' aesthetic is based on the notion that there is some intrinsic goodness in perception, neither captured nor perverted by the aestheticization process. (p. 26)

Before attempting to recruit operators and visitors interested in considering a circumpolar imaginary, the *Forum on the Sub-Antarctic* needs to convince the world that the concept contains a "good aesthetic" and that the region has not simply been aestheticized. Establishing what constitutes that "intrinsic goodness" clearly is an ethical question; deciding whether or not it has been neither "captured nor perverted" by the process of aestheticization, the production of the sub-Antarctic clearly is a political question. For the sub-Antarctic expedition cruise tourist, as subject, experiencing the material elements of *Solander* requires a certain distance to be observed between subject and object, the thing available to be experienced. This being distanced from the object is what allows the aestheticization of the object.

> Distance and proximity are aestheticized terms. They imply a perceiving subject and a perceived object. They are part of Immanuel Kant's language of aesthetics- in order to have aesthetic appreciation you have to have an appropriate distance towards the aesthetic 'thing.' (Morton 2007, p. 28)

4.7 Conclusion

One of the implications of establishing the sub-Antarctic region as an entity independent from Antarctica, effectively a process of reification, is the inevitability of aestheticization. Currently, as an imaginary, the region is free from the ideological rigidity that an aesthetic both imposes and reflects. I have deliberately avoided any attempt at untangling possible imaginaries involving those islands associated with the Antarctic Peninsula; that is too complex a task. Immediately, *Shackleton* springs to mind, quickly to be rejected as only an appendage to the narratives of Antarctica, the very situation that is to be avoided. Considering only those islands associated with New Zealand and Australia has been illustrative of the challenges a circumpolar focus brings. Shelton (2011) identified two research agendas for sub-Antarctic tourism. One of these involves visitors' experiences. As far as achieving the aims of the *Forum on the Sub-Antarctic* is concerned, the next step must be a series of empirical studies to establish how tourists perceive the sub-Antarctic region now and what would be required for them to view it more as the *Forum* would wish.

References

Antarctic Tasmania (2011) 2011 3rd international forum on the sub-Antarctic. Retrieved 2 Oct, 2011, from http://www.sub-antarctic.org/docs/Program2011.pdf

Ashcroft B (2001) Post-colonial transformation. Routledge, London

Aurora Expeditions (2011) Mawson's centenary celebration overview. Retrieved 24 Feb, 2011, from http://www.auroraexpeditions.com.au/expedition/mawsons-centenary-celebration

Baldacchino G (2006) Islands, island studies, island studies journal. Island Stud J 1(1):3–18

Bendix R (2002) Capitalizing on memories past, present and future: observations on the intertwining of tourism and narration. Anthropol Theory 2(4):18

Bertram E, Muir S, Stonehouse B (2007) Gateway ports in the development of Antarctic tourism. In: Snyder J, Stonehouse B (eds) Prospects for polar tourism. CABI, Wallingford, pp 123–146

Blaikie A (2005) Imagined landscapes of age and identity. In: Andrews GJ, Phillips DR (eds) Ageing and place: perspectives, policy, practice. Routledge, London

Boyer P, Haywood E (2006) Islands of the southern ocean: Information for delegates to the international forum on the sub-Antarctic, Hobart, July 2006. CCAMLR, Hobart

Cartier C (2005) Introduction: touristed landscapes/seductions of place. In: Cartier C, Lew A (eds) Seductions of place: geographical perspectives on globalization and touristed landscapes. Routledge, London

Cessford G, Dingwall P (1998) Research on ship-borne tourism in the Ross Sea region and the New Zealand sub-Antarctic islands. Polar Rec 34(189):7

Chaudhuri U (1995) Staging place: the geography of modern drama. University of Michigan Press, Ann Arbor

Crouch D, Desforges L (2003) The sensuous in the tourism encounter. Introduction: the power of the body in tourism studies. Tour Stud 3(1):5–22

Crouch D, Aronsson L, Wahlstrom L (2001) Tourist encounters. Tour Stud 1(3):253–270

Dann GMS, Jacobsen JKS (2003) Tourism smellscapes. Tour Geogr 5(1):3–25

Davies B, Harre R (1999) Positioning the discursive production of selves. J Theory Soc Behav 20(1):20

Department of Economic Development, Tourism and the Arts (2011) Developing Tasmania's Antarctic sector. Retrieved 5 Nov, 2011, from http://www.development.tas.gov.au/?a=50937

Edensor T (2001) Performing tourism, staging tourism: (re)producing tourist space and practice. Tour Stud 1(1):22

Escott-Inman H, Prater E (1911) The castaways of disappointment island: being an account of their sufferings. S.W. Partridge & Co, London

Fell D (2002) Campbell Island: land of the blue sunflower. David Bateman, Auckland

Fensham R (2000) Foreword. In: Mansfield N (ed) Subjectivity: theories of the self from Freud to Haraway. New York University Press, New York, pp v–vi

Gergen KJ (1991) The saturated self: dilemmas of identity in contemporary life. Basic Books, New York

Gibbs G (2006) Ghosts of Gondwana: the history of life in New Zealand. Craig Potton Publishing, Nelson

Green MC (2005) Transportation into narrative worlds. In: Tesser A, Wood JV, Stapel DA (eds) On building, defending and regulating the self: a psychological perspective. Psychology Press, New York, pp 53–76

Hall CM, Müller DK, Saarinen J (2009) Nordic tourism: issues and cases. Channel View Publications, Bristol

Halpenny E (2001) Islands and coasts. In: Weaver D (ed) The encyclopedia of ecotourism. CABI, Wallingford

Hartley G (2003) The abyss of representation: Marxism and the postmodern. Duke University Press, Durham

Hatton S, McGurcan M, Wang X-J (2010) Keg party extreme and conversation party: two multimodal interactive narratives developed for the small lab. In: Page R (ed) New perspectives on narrative and multimodality. Routledge, New York, pp 202–216

Hollinshead K (2009) Review – tradition and the declarative reach of tourism: recognizing transnationality – the articulation of dynamic aboriginal being. Tour Anal 14:537–555

Hull J (2011) Sub-Arctic tourism: sub-Arctic tourism introduction. In: Maher P, Stewart E, Lück M (eds) Polar tourism: human, environmental and governance dimensions. Cognizant Communication Corporation, New York, pp 187–205

Jackson A (2006) International instruments and arrangements in the sub-Antarctic. Paper presented at the international forum on the sub-Antarctic, Hobart, Tasmania, 6–7 July 2006

Jamal T, Hill S (2002) (Post)touristic spaces of (in)authenticity. In: Dann G (ed) The tourist as a metaphor of the social world. CABI, New York

Jaworski A, Pritchard A (2005) Discourse, communication and tourism. Channel View Publications, Clevedon

Lanchester J (2011) Short cuts. London Rev Books 33:1

Landau D (2007) Tourism in polar regions and the sub-Antarctic islands. Paper presented at the international forum on the sub-Antarctic, Hobart, 1–2 August 2007

Law J (2000) On the subject of the object. Configurations 8(1):29

Lewis J, Landreth R (2010) Scientists to visit island. Otago Daily Times, 3 February, p 1

Liggett D (2011) The sub-Antarctic as an economic and environmental asset. Paper presented at the 3rd international forum on the sub-Antarctic, Hobart, Tasmania, 1–2 August 2011

Maher P, Steel G, McIntosh A (2001) Antarctica: tourism, wilderness and 'ambassadorship'? Paper presented at the Science and stewardship to protect and sustain wilderness values. Seventh world wilderness congress symposium, Port Elizabeth, South Africa, 2–8 November 2001

McCabe S (2009) Who is a tourist? Conceptual and theoretical developments. In: Tribe J (ed) Philosophical issues in tourism. Channel View Publications, Bristol, pp 25–42

Mels T (2008) Analysing environmental discourses and representations. In: Anderson K, Braun B (eds) Environment: critical essays in human geography. Ashgate, Aldershot, pp 385–399

Miller D (2011) Opening address. Paper presented at the 3rd international forum on the sub-Antarctic: the future of the sub-Antarctic region: its global significance and value, Hobart, Australia, 1–2 August 2011

Morton T (2007) Ecology without nature: rethinking environmental aesthetics. Harvard University Press, Cambridge, MA

Page R (2010) Introduction. In: Page R (ed) New perspectives on narrative and multimodality. Routledge, New York

Peat N (2003) Subantarctic New Zealand: a rare heritage. Department of Conservation, Invercargill

Perkins HC, Thorns DC (2001) Gazing or performing? Reflections on Urry's tourist gaze in the context of contemporary experience in the antipodes. Int Sociol 16(2):185–204

Quasthoff U (1997) An interactive approach to narrative development. In: Bamberg M (ed) Narrative development: six approaches. Lawrence Erlbaum Associates, Mahwah, pp 51–83

Rappaport J (1993) Narrative studies, personal stories, and identity transformation in the mutual help context. J Appl Behav Sci 29(2):239–255
Rubinstein DH (1992) Pacific history: papers from the 8th Pacific History Association conference. University of Guam Press & Micronesian Area Research Center, Mangilao
Russ R (2007) History, exploration, settlement and past use of the sub-Antarctic. Paper presented at the international forum on the sub-Antarctic, Hobart, 6–7 July 2006
Ryan M (2007) Toward a definition of narrative. In: Herman D (ed) The Cambridge companion to narrative. Cambridge University Press, Cambridge, pp 22–35
Seigel J (2005) The idea of the self: thought and experience in Western Europe since the seventeenth century. Cambridge University Press, Cambridge
Selkirk P, Quilty P, Davies M (2007) Preface. Paper presented at the forum on the Sub-Antarctic: The future of the sub-Antarctic region: Its global significance and value, Hobart, 6–7 July 2006
Shelton EJ (2007) Evaporating the essence: towards an understanding of sub-Antarctic visitation. Paper presented at the 5th international coastal and marine tourism congress: balancing marine tourism, development and sustainability, Auckland, New Zealand, 11–15 September 2007
Shelton EJ (2011) Sub-polar tourism: Introduction; sub-Antarctic. In: Maher P, Stewart E, Luck M (eds) Polar tourism: human, environmental and governance dimensions. Cognizant Communications, New York
Somers MR (1994) The narrative constitution of identity: a relational and network approach. Theory Soc 23(5):605–649
Stewart P (1984) Benediction and sea longing. Am Poet Rev 13(5):1
Strong T (2000) Review of James Holstein and Jaber Gubrium's the self we live by: narrative identity in a postmodern world. Gecko J Deconstr Narrat Pract 5(3):81–85
Vlachou A (2001) The promise of finance: banks and community development. In: Gibson-Graham JG (ed) Re/presenting class: essays in postmodern Marxism. Duke University Press, Durham
Wahrman D (2004) The making of the modern self: identity and culture in the eighteenth-century England. Yale University Press, New Haven
Walton D, Dingwall P (1995) Foreword: determining the priorities for conservation management of subantarctic islands. In: Dingwall P (ed) Progress in conservation of the Subantarctic islands. IUCN, Gland, pp ix–x

Chapter 5
Reaching Consensus on Antarctic Tourism Regulation

Calibrating the Human-Nature Relationship?

Kees Bastmeijer and Machiel Lamers

Abstract The states that jointly manage Antarctica were not yet able to reach consensus on a number of specific Antarctic tourism management issues. While recognizing that many factors influence the international decision-making process, this chapter aims to explore one rarely discussed philosophical factor that may be of great significance for understanding the difficulties in reaching consensus: the possible differences in the human-nature relationship among the Consultative Parties (CPs) and expert organizations. Based on an explorative analysis of relevant documents, it is concluded that CPs and expert organizations consistently reject the "mastery" attitude, both in general Antarctic policy instruments and in tourism-specific documents. The relevant documents relating to two subthemes (the acceptability of permanent land-based tourism facilities and large-scale adventure or sporting events) show that there is a strong tendency to use "rational arguments" that do not reflect human-nature relationships; however, a closer look reveals that underneath these relatively neutral positions, substantially different human-nature attitudes appear to be hidden. These differences may not block consensus regarding general policy statements on Antarctic tourism, as these statements leave sufficient space for different interpretations; however, different attitudes towards nature may well constitute a hurdle in reaching consensus on concrete management issues. It is most likely that CPs with different human-nature relationships have different views on what the specific "values" of Antarctica are and how these values could best be protected, and, consequently, it is also most likely that these CPs have different

K. Bastmeijer, Ph.D.
Tilburg Law School, Department European and International Public Law, Tilburg University, PO Box 90153, 5000 LE Tilburg, The Netherlands
e-mail: c.j.bastmeijer@uvt.nl

M. Lamers, Ph.D. (✉)
Environmental Policy Group, Wageningen University, PO Box 9101, 6700 HB Wageningen, The Netherlands
e-mail: machiel.lamers@wur.nl

D.K. Müller et al. (eds.), *New Issues in Polar Tourism: Communities, Environments, Politics*, DOI 10.1007/978-94-007-5884-1_5,

opinions on what norms should be set in respect of specific tourism developments. This makes it understandable that the approach of the ATCM to focus strongly on (the desirability of) norm setting is not always successful. Underlining the explorative nature of this study, the authors would recommend further research on human-nature relationships in the Antarctic tourism context. More knowledge and consciousness of differences in human-nature attitudes might invite stakeholders to search for management solutions based on a greater understanding of each other's convictions.

Keywords Antarctica • Tourism • Regulation • Future vision • Human-nature relationship

5.1 Introduction

Antarctica's isolation from human settlements has provided an effective protection from intensive human visitation for a very long time. However, in the past decades, human activities in Antarctica – in particular tourist activities – have grown and diversified rapidly. Since the early 1990s, the number of tourists visiting Antarctica has grown substantially, particularly in the Antarctic Peninsula region. The total number of tourists taking part in Antarctic travel itineraries has increased from around 2,500 in 1990 to more than 45,000 in 2008 and decreased to 35,000 in 2010 as a result of the global economic crisis. More than 90% of these tourists were ship-borne, but Antarctic tourism is becoming more and more diverse. The traditional expedition cruises, involving small- to medium-sized ships, rubber boat landings, and educational programs, have been complemented with large cruise liners making no landings, overflights, fly-sail operations, as well as forms of land-based tourism using aircraft for transportation. In expedition cruises and land-based itineraries, an increasing range of adventurous activities are offered, including helicopter excursions, kayaking, overnight camping, scuba diving, mountain climbing, cross-country skiing, and marathons (Bastmeijer and Roura 2004; Stonehouse and Crosbie 1995; Lamers and Gelter 2011).

The states that collectively manage Antarctica under the Antarctic Treaty System (ATS) have been well aware of this development and of the related concerns regarding the safety of tourists, the interaction between science operations and tourism, and the potential direct, indirect, or cumulative effects on Antarctic ecosystems and environmental values. During the last two decades, the Consultative Parties (CPs) to the Antarctic Treaty adopted several instruments to regulate Antarctic tourism (see Sect. 5.2). The governmental system of the ATS is paralleled by a comprehensive self-regulation system, developed by the "International Association of Antarctica Tour Operators" (IAATO), an industry association founded in 1991. The "Antarctic and Southern Ocean Coalition" (ASOC) represents the international environmental movement and forms another important expert group that is making efforts to influence tourism policy decision making. The ATS instruments, applicable to

Antarctic tourism activities, constitute quite a comprehensive regulatory system, but it is also clear that a substantial number of management questions relating to Antarctic tourism did not yet receive an answer (e.g., Bastmeijer and Roura 2004; Bastmeijer 2011; Netherlands and United Kingdom 2011). For instance, uncertainty exists on how the Antarctic Treaty Consultative Meeting (ATCM), the annual meeting of the Consultative Parties, should respond to a possible future increase of vessels, used for tourism purposes, sailing the flag of states that are not Contracting Parties to the Antarctic Treaty and the Protocol. It is also unclear whether the Consultative Parties are willing to establish an international observer system to improve supervision of tourism activities in Antarctica. And, although intensively debated for about 5 years, the ATCM has not yet reached consensus on the desirability of additional regulations (e.g., a prohibition) regarding the establishment of permanent facilities for tourism in Antarctica (such as hotels) (Bastmeijer et al. 2008). Similarly, the question of whether Antarctica should be open for all forms of tourism activities (from hiking and kayaking, to the use of motorized vehicles and large-scale sport events) has not received a clear answer from the ATCM. Consensus has also not been reached on the question whether the prevention of cumulative impacts by the continuing increase of visitation should require the adoption of more strategic instruments, such as a zoning system or the establishment of a cap on visitor numbers per Antarctic region or site.

Why is it so difficult to reach consensus on these and several other tourism management questions at the ATCM? The answer is complex as many different factors influence the process of international negotiations and decision making. For the ATCM, such factors include the number of CPs (28) that only take decisions on the basis of full consensus; the territorial claims by seven CPs, whose claims have been "frozen" with Article IV of the Antarctic Treaty but still influence the positions of CPs; the differences in domestic law of CPs; the different levels of expertise of CP representatives at the ATCMs; the limited time available at ATCMs for reaching consensus; the lack of a strategic approach in discussing tourism issues over the last two decades; and the different views on how serious the concerns are and the appropriateness of policy responses. The relevance of these factors may differ from issue to issue and from year to year.

The aim of this chapter is not to discuss the complexity of the decision-making process and all relevant factors that may have an influence (e.g., Bastmeijer and Roura 2004; Enzenbacher 2007). This chapter aims to explore one possible other, more philosophical, factor that has not been discussed in an Antarctic context before, but that may be of great significance for understanding the decision-making process within the ATCM: the possible differences in orientation of the human-nature relationship among the CPs and expert organizations. It is our assumption that such basic attitudes towards nature may influence the decision-making process, for instance because these attitudes may affect the views regarding the values of Antarctica that require protection, the seriousness of effects that human activities have on such values, and the types of measures needed to limit these effects. The aim of this chapter is reflected in the following two research questions that structure this chapter:

- Which type(s) of relationship between humans and nature – as described by environmental philosophers – can be identified in legally binding and soft-law instruments of the ATS and in ATCM papers and reports, relating to tourism in Antarctica?
- If there are any differences in human-nature attitudes among the Consultative Parties and expert organizations, are there any indications that these differences influence the decision-making process at the ATCMs?

This chapter makes a start in answering these questions based on a small-scale and explorative analysis of the relevant documents. The main source of data analyzed in this study are the final reports of the ATCMs of the last decade, the recommendations and measures adopted at these ATCMs, and the information papers (IPs) and working papers (WPs) on tourism related issues, tabled by the CPs, IAATO, and ASOC at these ATCMs. We have made use of the Antarctic Treaty Secretariat's document database, in which these documents have been collected and made publicly available (AT Secretariat 2011). Selection of relevant documents was facilitated by the search engine of the database that allows for selection based on dates, keywords, and authors. The resulting collection of documents was qualitatively analyzed using a typology of human-nature relationships (basic attitudes), distinguished by contemporary environmental philosophers, that is, the master, the steward, the partner, and the participant (Zweers 1995). Recognition of these attitudes in the various documents were noted, further analyzed, and provided with illustrative citations. The methodology used is limited in the sense that the results only represent observations and reflections made by the authors; representatives of CPs and expert organizations have not been interviewed directly in respect of human-nature attitudes. Nonetheless, the authors believe that the results derived through this methodology do shed some light on a fundamental dimension of Antarctic policy making that has so far lingered in the background.

After a brief introduction on Antarctic tourism regulation in Sect. 5.2, a brief historical introduction of dominant views on human-nature relationships in Western societies is given, as well as the main attitudes that have been put forward for this relationship. A typology of four human-nature relationships will be introduced in Sect. 5.4. In Sect. 5.5 the question to what extent these human-nature relationships can be recognized in the history and legal base of the Antarctic Treaty System discussed. Section 5.6 presents an analysis of various positions of CPs and expert organizations regarding tourism management in Antarctica based on the human-nature relationships. Section 5.7 concludes to what degree these different positions influence Antarctic tourism policy decision making.

5.2 Regulating Antarctic Tourism: A Brief Introduction

The most important instrument regulating tourism in Antarctica is the Protocol on Environmental Protection to the Antarctic Treaty in 1991 (Protocol 1991), which entered into force on 15 January 1998. This Protocol establishes a comprehensive system of obligations and prohibitions, addressing most human activities, including all types of tourism activities, in the region south of 60° south latitude. Consequently, tourist activities are subjected to the Protocol's provisions on environmental impact assessment, the protection of Antarctic flora and fauna, waste disposal, the protection of Antarctic Specially Protected Areas and historic sites, and regulations on liability for damage resulting from environmental emergencies. The CPs must ensure that the provisions of the Protocol are implemented in their domestic legal and administrative systems and applied in practice to all Antarctic activities under their jurisdiction (Bastmeijer 2003). At the ATCMs, the CPs discuss the (legal and practical) implementation of the Protocol and additional instruments as well as the need to adopt new international management measures. Since the adoption of the Protocol, the ATCM adopted several additional recommendations (soft law) and measures (legally binding after approval) on Antarctic tourism. Some of these instruments will be discussed in more detail below.

In parallel to this governmental system, IAATO has imposed a wide range of operational procedures and environmental standards upon its member companies (Splettstoesser 2000). IAATO membership has grown from seven members in 1991 to more than 100 members in 2011 in different membership categories (IAATO 2011). Currently, the vast majority of the tourist activities in Antarctica are organized by IAATO member companies; however, it is not self-evident that this practice will be continued (Haase et al. 2009).

5.3 Human-Nature Relationships: A Brief Historic Perspective

It has been stated that the environmental degradation in the past has mainly been the result of the dominant attitude of man against nature and the earth. In a seminal paper on the historical roots of our ecologic crisis in Science, Lynn White Jr. attributed this dominant attitude of humans to nature to the Judeo-Christian tradition (White 1967):

> Man named all the animals, thus establishing his dominance over them. God planned all of this explicitly for man's benefit and rule: no item in the physical creation had any purpose save to serve man's purposes. And, although man's body is made of clay, he is not simply part of nature: he is made in God's image. (White 1967: 4)

To support this view, one could refer to texts from the bible, in particular from Genesis 1:28 and Genesis 9:2–3 (Thomas 1983). In Lynn White's opinion, the dominant attitude led to an ecologic crisis mainly as a result of the scientific and industrial revolutions. Science and technology "joined to give mankind powers which, to judge by many of the ecologic effects, are out of control. If so, Christianity bears a huge burden of guilt" (White 1967: 5). Cliteur also observes that the subordinate position of animals in Western culture may be explained from the dominance of the Judeo-Christian world image, allied to a dominant line in ancient

Greek philosophy: the rationalist tradition (Cliteur 2005). During recent decades, White's opinion has been partly or wholly endorsed by some, but also heavily criticized by others (e.g., Minteer and Manning 2005; Harrison 1999). Some authors point to other interpretations of biblical texts which in fact emphasize man's responsibility towards nature as a steward (Chen 1995). Others point to historical examples of ecologically irresponsible behavior of societies without a Judeo-Christian tradition (Minteer and Manning 2005; Diamond 1992, 2004). Some authors go far back in time in their search for explanations for the dominant human attitude and argue that the distance between man and nature dates from the rise of agriculture (Wells 2010) and the domestication of animals. In that period, the relationship with nature changed and human dominance over nature was enhanced. From the start of domestication, humankind has been able to influence evolution to its own advantage.

The limited space available does not allow more comprehensive attention for this debate as part of this article; however, it is clear from the rich historic literature that the idea that nature is meant to serve mankind has been a strong conviction in the Western world for more than 2,000 years. This conviction is clearly reflected in documents written by lawyers and philosophers from ancient Greek (e.g., Aristotle), Roman (e.g., Cicero), medieval (e.g., Aquinas), and enlightened times (e.g., Grotius and Locke). Certainly, contrasting views can also be found. The Roman poet Lucretius (99–55 BCE) (Schrijvers 2008) and, much later, Spinoza (1632–1677) refused to accept that the earth exists solely for the benefit of humankind (Steenbakker 1999), but their discussions confirm that their opinions were not shared by many.

5.4 Types of Human-Nature Relationships in Environmental Philosophy

In the 1970s and 1980s, numerous environmental philosophers have developed classifications of human-nature relationships, or "basic attitudes" (De Groot et al. 2011). Although there are differences in descriptions of these types, as well as a range of subcategories, four "basic attitudes" are identified by most authors: the master, the steward, the partner, and the participant. In this chapter, we will follow these categories as defined by the Dutch environmental philosopher Wim Zweers (1995).

The *master* involves an image that regards nature as subordinate to man. Humans may use nature and adjust it for their own purposes and benefit. If humans experience problems as a result of these adjustments, the conviction is that technology will provide a solution. Growth of human society and economy is paramount. Ecological awareness is valuable, but mainly to learn how we may use nature more efficiently for our own benefit (Zweers 1995). In the attitude of the master, man dominates nature and is not accountable to anyone.

The attitude of the *steward* is also anthropocentric in character and considers nature largely as a resource for man to use. The difference with mastery is that, according to the steward, man also bears a certain responsibility for nature. This responsibility may exist towards God (Christian stewardship) or towards other people, for example, regarding future generations of people (secular stewardship) (Zweers 1995: 39). The problem with stewardship is that it remains unclear what God, other people, or future generations expect the steward to do. Does God want the steward to polish up nature and perfect it, or does he wish to preserve what he once created? How do we know what future generations want or need, and can we anticipate possible future technological developments? This lack of certainty gives the steward much space for interpretation. In Zweers' opinion, the steward is somewhat conservative (Zweers 1995: 42). This means making use of nature while taking into account its carrying capacity: man may use 'the interest' for his own benefit, but he must not touch the capital so as not to deprive future generations from the interest accruing to them. In the view of the steward a balance must be struck between justified personal use of nature (anthropocentric attitude) and responsibility for nature.

Unlike the relationships described above, the *partner* and *participant* attribute a separate, intrinsic value to nature. The value of nature and the importance of nature conservation is recognized, independent from the direct usefulness of nature for humans. Both relationships may therefore be called ecocentric (as opposed to the above-mentioned anthropocentric relationships). The *partner* places himself on one level with nature (man and nature are on a par), while the *participant* mainly places himself in nature (man is part of nature). Zweers argues that, in a certain sense, the partner has no internal relation with nature but holds freedom of choice to be himself separate from nature. Partnership with nature is something extra, which he may or may not enter into (Zweers 1995: 49). This is different for the participant who considers humans as partaking in an event that includes him, that transcends him, that is greater, and that will go on if he were not there (Zweers 1995: 52). In short, the partner wants to cooperate with nature for mutual benefit; the participant wants to experience nature.

Table 5.1 summarizes the main characteristics of the four human-nature relationships as briefly described above. The table also describes possible manifestations of these attitudes in Antarctic tourism policy positions (elaborated in Sects. 5.5 and 5.6).

5.5 Human-Nature Relationships and the Antarctic Treaty System

The history of economic activities in the Antarctic region clearly shows that there have been some clear periods of "mastery" of man over nature in the Antarctic and sub-Antarctic region (Headland 2009; Basberg 2005). It is estimated that within one decade (from 1819 to 1827), more than 800,000 skins of fur seals have been taken

Table 5.1 "Characteristics" of the human-nature attitudes (Van den Born 2007; Zweers 1995) and possible manifestations in Antarctic tourism policy views

Attitude	General characteristics	Possible manifestations in Antarctic tourism policy views
Master	Strongly anthropocentric Nature is meant for the use by humankind; nature is subordinated to man and man controls nature. Humankind may alter nature and nature conservation cannot justify limitations of human ambitions. Value of nature is determined by the use value for humankind. Use may be unlimited; if problems would arise (problems for mankind), there is a conviction that solutions will come from new technology developments. Humankind is not accountable	Antarctica is meant for human's enjoyment. Antarctica should be open for anyone's use. Antarctica as a playground and building ground; nature conservation cannot justify prohibitions of hotels or types of tourist activities. If there would be any concern in respect of certain types of tourism, problems should be resolved by technology. Tourism may be a way to sustain or strengthen territorial claims
Steward	Anthropocentric Nature is meant for the use by humankind; nature is subordinated to man and man controls nature, but: Use is not unlimited: overuse would be in conflict with the will of God or with the needs of future generations of mankind (sustainable use/wise use; use value of nature's rent). Thus, (a) extinction of species must be prevented and (b) options for future generations to enjoy nature must be safeguarded. A balance must be found between personal use and responsibility	Humankind may enjoy Antarctica's beauty; Tourism creates ambassadors (stewards) for Antarctic conservation. All types of tourism are allowable, also if these activities alter Antarctica, as long as: (a) there are no threats of extinction of species and (b) options for future generations to enjoy Antarctica are not diminished
Partner	Dualistic Humankind and nature are equal and exist as separate entities. Humankind and nature may decide to collaborate in partnership, but this is not necessary (freedom of choice). Intrinsic value of nature is recognized	It would be best to allow all those types of tourism that provide enjoyment to the tourists and at the same time promote Antarctica's protection (education through tourism). Human safety and environmental protection go hand in hand. Visiting Antarctica is a once in a lifetime and a life-changing opportunity

(continued)

Table 5.1 (continued)

Attitude	General characteristics	Possible manifestations in Antarctic tourism policy views
Participant	Ecocentric Humankind is part of nature; nature transcends humankind, and therefore. humankind must obey the rules of nature. Humankind should not alter nature (preservation). It is best to enjoy nature by experiencing nature in its original condition (wilderness experience)	Humans are relatively alien to the Antarctic ecosystem. Antarctica, as one of world's last unspoiled wildernesses, does offer a nature experience in its most pure way. Tourists must behave in Antarctica in full accordance with nature's rule. Only those types of tourism that do not alter Antarctica in any way should be considered acceptable

around the South Shetland Islands (Pearson and Stehberg 2006: 337). Whaling in the Antarctic region started later (beginning of the twentieth century) but was also characterized by mastery. Technological innovations, of the pelagic factory ships combined with catcher fleets, supported this attitude (Holdgate 1987: 130). Later in the twentieth century, there have been examples of overexploitation of some Southern Ocean fish stocks (Parkes 2000: 83).

One could argue that stewardship could only emerge after issues of sovereignty had been settled or regulated (Heap 1990). This argument is supported by observing the Antarctic Treaty and subsequent legal instruments adopted by the ATS, as providing safeguards against mastery. With Article 4 of the Treaty, the territorial claims are "frozen," while at the same time, human activities are regulated on the basis of consensus among the CPs (AT 1959; Hayton 1960). Avoidance of overexploitation of natural resources has been a central aim of ATS instruments since the beginning of the 1960s. Already at the first ATCM, the representatives recommended to their governments that "they recognize the urgent need for measures to conserve the living resources of the Treaty area and to protect them from uncontrolled destruction or interference by man" (Vicuña 1996: 128). The 1972 "Convention for the Conservation of Antarctic Seals" explicitly recognizes that seal populations "should not be depleted by overexploitation, and hence that any harvesting should be regulated so as not to exceed the levels of the optimum sustainable yield" (CCAS 1972). The 1980 "Convention on the Conservation of Antarctic Marine Living Resources" also established a comprehensive regime that regulates harvesting activities to ensure "sustained conservation of Antarctic marine living resources" (CCAMLR 1980).

The aim of reducing significant adverse impacts by human activities in Antarctica was a fundamental principle for developing the Convention on the Regulation of Antarctic Mineral Resource Activities (CRAMRA 1988). Beeby explains that the main aim of CRAMRA was to fill the institutional void, which gave rise to past examples of mastery in the Antarctic region (Beeby 1990). This is also a characteristic of the Seals Convention and several other ATS instruments: taking a proactive approach, regulations are adopted in respect of activities before such activities would develop, in order to prevent "no-rule situations."

In 1989 it became clear that CRAMRA would never enter into force (Blay 1992: 378). Instead, in 1991 the Protocol on Environmental Protection to the Antarctic Treaty was signed, a comprehensive system of environmental protection that entered into force in 1998. The Protocol clearly aims to prevent mastery in the Antarctic Treaty area, through the provisions of Annex II regarding the protection of Antarctic flora and fauna, the provisions on waste management in Annex III, and the prohibition of mineral resources activities for other reasons than scientific research purposes in Article 7. This last prohibition can be reviewed in 2048 in accordance with the requirements set out in Article 25 of the Protocol (Protocol 1991).

Since the adoption of the Protocol, the ATS has experienced new challenges, such as the substantive increase of tourism activities, both in numbers and in diversity. At the ATCMs since 2004, issues related to Antarctic tourism were intensively debated in the Working Group on Tourism and Non-Governmental Activities (Bastmeijer 2011).

5.6 Human-Nature Relationships and the Regulation of Antarctic Tourism

5.6.1 Consensus at the General Policy Level on the Rejection of "Mastery" in Antarctic Tourism

Already during the negotiations of the Protocol (1990–1991), concerns were expressed about the development of tourism and other nongovernmental activities in Antarctica. Consensus could not be reached on the need for a separate annex on tourism; however, tourism received substantial attention during the ATCMs at the beginning of the 1990s. These discussions make clear that the ATCM considered tourism developments that would reflect "mastery" as unacceptable. For instance, in the tourism guidelines, adopted in 1994, the CPs reiterated that "the Environmental Protocol designates Antarctica as a natural reserve devoted to peace and science" and that "the Protocol seeks to ensure that human activities, including tourism, do not have adverse impacts on the Antarctic environment, nor on its scientific and aesthetic values" (ATCM 1994: 35–45).

This position of the CPs was paralleled by a similar position of the Antarctic tourism sector. From the start of its establishment in 1991, IAATO has been advocating responsible tourism that would result in "taking only photographs" and "leaving only footprints." IAATO has been very successful in showing the ATCM its value in promoting responsible tourism. One of the consequences of this active role and successful communication was that during the 1990s proposals for additional regulations of tourism received little support by CPs. The view that tourism was sufficiently regulated by the Protocol and IAATO dominated at ATCMs.

Particularly since 2004, more and more questions arose on whether this approach was sufficient and whether additional regulatory initiatives of the ATCM were required. ASOC expressed concerns that Antarctic tourism could well take shapes that would have the characteristics of mastery. For instance, ASOC stated that tourism may in the course of the years "appropriate" certain sites for tourism purposes:

> Once a site becomes a well-established destination, tour operators will in all probability use it for the foreseeable future. As a result there is a risk of permanent appropriation of certain sites for tourism purposes, sometimes concurrently with other actual or potential activities at the site. (ASOC 2006a: 4)

According to ASOC, "[t]he industry now argues de facto rights of access, and often priority of access to sites it has used for some time" (ASOC 2006b: 5). This issue of "appropriation of sites" has not seriously been discussed at the ATCMs.

IAATO would most likely not agree with this view. At various occasions, the association has stressed that the impact of tourism in Antarctica is limited and that sites are effectively managed through existing instruments, particularly the bylaws and guidelines adopted by IAATO itself. For instance, in respect of the use of sites for tourism, IAATO stated in 2004:

> In 35 years of Antarctic tourism there is very little discernible impact from tourist activities at any of the landing sites in the Antarctic. The few examples include a minor gravel walking trail from the beach up to Neptunes Window at Deception Island, and a trail at Penguin Island. All other landing sites exhibit mere transient footprints in the snow. Rarely, if at all, are items discarded by tourists found ashore at any of the sites. This is mainly because of the numerous guidelines and operational procedures IAATO have put in place over the last 14 years to minimize possible environmental impacts." (IAATO 2004: 4)

However, particularly the last few years, the CPs expressed more explicitly the "own responsibility" of the ATCM to regulate Antarctic tourism effectively. For instance, in 2007, the USA stated:

> It should be acknowledged that [IAATO] has done an admirable job of implementing rules for its members that promote safety and environmental stewardship. Nevertheless, the overall responsibility for the management of all human activities in Antarctica lies with the ATCP's. (...) Significant increases in tourism are placing pressure on the Antarctic environment that requires close monitoring and evaluation. (USA 2007: 3)

At the same ATCM, Resolution 5 (2007a, b) was adopted, containing the general policy statement that the Parties "discourage any tourism activities which may substantially contribute to the long-term degradation of the Antarctic environment and its dependent and associated ecosystems" (ATCM 2007b), again, a recent rejection of mastery in the context of Antarctic tourism.

That the ATCM explicitly rejects the attitude of mastery is even more clearly reflected in the "general principles" for Antarctic tourism, adopted with Resolution 7 (2009a, b). The aim of this resolution is "to inform and guide further work in managing Antarctic tourism activities." One of these principles states:

> Tourism should not be allowed to contribute to the long-term degradation of the Antarctic environment and its dependent and associated ecosystems, or the intrinsic natural wilderness and historical values of Antarctica. In the absence of adequate information about potential impacts, decisions on tourism should be based on a pragmatic and precautionary approach, that also incorporates an evaluation of risks. (ATCM 2009a)

Similar to the previous discussed instruments, this resolution clearly reflects the CPs policy aim to prevent tourism developments in Antarctica from having characteristics of overexploitation. While the "responsibility" that is reflected in the resolution fits well to the attitude of stewardship, this resolution contains also wording that reflect the more ecocentric attitudes of partner and participant ("intrinsic natural wilderness and historical values of Antarctica").

5.6.2 A Closer Look: No Consensus Regarding Several Concrete Issues

In view of this consensus regarding the rejection of "mastery" through Antarctic tourism, one would expect little difficulties in respect of decision making on specific issues; however, there are various issues, related to Antarctic tourism, that have been the subject of lengthy debates while consensus could not be reached, possibly due to differences in human-nature relationships. To illustrate this, two subthemes are discussed below.

5.6.2.1 Permanent Facilities for Tourism in Antarctica

One of these issues, which has been the subject of debate since 2004, is the question whether permanent facilities and infrastructures for tourism should be allowed to develop on the Antarctic continent or Antarctic islands. A range of criticisms were formulated against the establishment of such facilities. Permanent facilities were considered inconsistent with the designation of Antarctica as a "natural reserve, devoted to peace and science" (New Zealand 2005), believed to "inadmissibly erode and restrict the research privilege" and contradict with the precautionary principle (Germany 2005). New Zealand also expressed concern that a shift of tourism activities from sea to land may cause difficult and sensitive discussions between states and other stakeholders on legal issues (sovereignty, jurisdiction, and private ownership), which may "severely test the ATS framework" (New Zealand 2005). According to France, the establishment of such facilities in Antarctica represents "activities whose environmental impact is far from insignificant (. . .), which justifies implementing regulations" (France 2006: 3). Finally, several CPs and expert organizations argued that land-based tourism facilities are inconsistent with the obligation to protect Antarctica's intrinsic aesthetic and wilderness values as laid down in the Protocol (e.g., New Zealand 2005; Australia 2005; ASOC 2008). For example, New Zealand states:

> By designating Antarctica as a "natural reserve" and by mandating the protection of its "wilderness values" it is apparent that the State Parties wished to preserve the continent as a wild, uninhabited area as far as possible. The creation of permanent or semi-permanent structures to accommodate numbers of people whose travel to Antarctica is primarily for the purposes of pleasure or adventure as opposed to science is very likely to infringe upon these values. (New Zealand 2005: 2)

In view of these arguments, many of the mentioned papers stressed that measures should be taken before it is too late, such as a prohibition on the establishment of permanent facilities for tourism on land.

However, some CPs were not convinced of the need to adopt such measures (ATCM 2005; 2006, 2007a). During discussions at the 2007 ATCM Argentina explicitly expressed that it "reserved its right to install at any time interpretative centres for tourists with some lodging capacity in any of its bases" (ATCM 2007a: 39, para.173). Others questioned the inconsistency of permanent facilities with the provisions of the Protocol, or the hierarchy between science and tourism. Some parties "believed that these activities could be controlled through existing instruments such as Environmental Impact Assessment procedures" (ATCM 2005). Various parties emphasized the unclear scope of the issue and the proposed management measures, particularly because terms such as "semipermanent," "facility," or "infrastructure" could become the subject of different interpretations among CPs. IAATO expressed the view that the development of serious proposals for such facilities should be considered unlikely. IAATO also argued that there are already structural facilities in Antarctica to support tourism and suggested that tourism may support ATCPs in funding the cost of scientific research and logistics.

All of these views clearly reject the mastery notion of establishing large-scale permanent facilities in Antarctica that would result in long-term degradation of the Antarctic environment. Most of the arguments, including the counter arguments, can be aligned with one of the other attitudes. Also Argentina would probably regard their position in the land-based tourism debate as not conflicting with the stewardship attitude. They might argue that education of tourists is important to support sustainable tourism and that the establishment of facilities in the direct vicinity of research stations has limited additional environmental consequences. Thus, at first sight, both sets of arguments hold on to rational and tangible aspects associated with this issue, such as the stipulations of Environmental Protocol, potential impacts, questions of definitions, real political international relations, and costs and benefits. Through this rational approach in debating, it appears that CPs are reluctant to show their human-nature relationships.

A closer look, however, makes clear that differences in attitude are reflected in the debates. For instance, the argument of the intrinsic values corresponds to the more ecocentric attitude of the partner and the participant. Probably based on the recognition of the high importance of the protection of wilderness values in Antarctica, Germany, France, and several other CPs (e.g., the Netherlands) are against the establishment of any hotel. However, Japan takes a different view:

> Japan stated that, under the Treaty and the Protocol and their domestic legislation implementing them, its government could discourage activities, including activities related to tourism, only when they have adverse environmental effect. According to Japan, a facility called "hotel" may or may not have those effects. (ATCM 2008: 53)

It may well be that these views – reflecting different human-nature relationships – receive all sufficient recognition in the more general policy statements, such as Resolution 5 (2007), in which CPs are recommended to "discourage any tourism activities which may substantially contribute to the long-term degradation of the Antarctic environment and its dependent and associated ecosystems" (ATCM 2007b). The vague and general terminology of such policy statements could receive the agreement of each CP, despite the different attitudes that seem to underlie opposing arguments expressed in the debate on the more specific issue of hotels.

5.6.2.2 Large-Scale Adventure or Sporting Events

Another example of a sub-debate that has not managed to reach consensus deals with the appropriateness and regulation of large-scale adventure or sporting events, such as marathons in Antarctica. At the 32nd ATCM, Chile "reiterating that the steady increase of large scale activities such as marathons on the Antarctic continent is disruptive to the function of scientists and other station activities in addition to posing a hazard to both tourists and the Antarctic environment" (ATCM 2009a, b: 56; Chile 2009). This sub-debate has spurred a range of different views. To start with, also in respect of this issue, various "rational" arguments were exchanged that do not clearly reflect any human-nature relationship. For instance, questions on the definition of the qualification of "large-scale" were raised (ATCM 2010: 70), and it was argued that more clarity on this issue would support competent authorities in assessing permit applications (ATCM 2009a, b: 57, 2010: 69).

However, similar to the previous subtheme (permanent facilities), a closer look at the debate shows that there appears to be substantial differences in attitude among the CPs that may at least in part explain why consensus could not be reached. Some CPs (e.g., Argentina, France, India, USA) have stated that proper implementation of existing regulations, such as EIAs, and proper planning should be sufficient for controlling negative effects of such activities (ATCM 2009a, b: 57, 2010: 69). Additional regulatory measures are not considered necessary. These views appear to resemble the attitude of the steward, in which all types of tourism activities are deemed appropriate as long as impacts to the environment and scientific activities can be minimized. Other CPs (e.g., The Netherlands, Germany) and ASOC have expressed the view that priority should be given to tourism activities that focus on educational enrichment and respect for the Antarctic environment and the intrinsic values of Antarctica, and can therefore not be carried out anywhere else (ATCM 2009a, b: 57, 2010: 69, 2011: 40). These views are closer to the partner and participant attitudes, in which appropriate tourist activities should be aligned with the wilderness and aesthetic values of Antarctica, which might require additional policy instruments and international coordination.

Possibly to prevent "subjectivity" in the debate and policy development in the ATCM, Australia and other CPs have stated that "tourism activities should be considered primarily in terms of their environmental impact" (ATCM 2010: 69). Similar to Resolution 5 (2007), CPs might reach agreement on such a statement; however, it would probably not be very helpful for the debate on this specific issue. For instance, as such a general statement does not specify the term "environmental impact" (e.g., are wilderness values to be taken into account?) and does not provide clarity on the threshold of what impact is considered acceptable, such a statement leaves room for all different views while concrete decision making is not possible.

5.7 Conclusions and Further Research

This chapter presents the results of an explorative analysis into different attitudes of the human-nature relationship (attitudes) in policy-making processes for Antarctic tourism. The limitations of the current analysis does not allow for drawing strong conclusions on the presence of a causal relation between differences in attitude and the failure to reach consensus. The aim of this chapter is more to initiate debate about the influence of such attitudes in Antarctic tourism policy-making processes as well as to explore the need for further research in this direction.

Based on the analysis, we can conclude that considering Antarctic tourism, the mastery attitude is generally and consistently rejected, both in general Antarctic policy instruments and in tourism-specific documents. CPs and expert organizations (ASOC and IAATO) clearly favor actions towards the protection of the Antarctic environment and natural resources; an attitude that fits in past and present discourse in which mastery is rejected in context of numerous Antarctic activities. However, a closer look at the Antarctic tourism policy documents does reveal the difficulty to reach consensus on several concrete sub-debates, such as on the acceptability of permanent land-based tourism facilities and large-scale adventure or sporting events. The influence of different human-nature relationships in these sub-debates is not easily recognized, since most statements of CPs and expert organizations have a neutral character, which do not correlate directly with one or more of the attitudes. Discussion papers and interventions of CPs at the ATCM tend to focus on tangible issues (e.g., interaction between actors, accidents) using rational arguments (e.g., financial costs, scientifically established impacts). Arguments relating to the more intangible issues (e.g., intrinsic values) and irrational arguments (e.g., "gut feelings") would likely reveal a greater diversity of attitudes, but CPs appear to be reluctant to use such arguments. There may be different explanations for this. For instance, attitudes of CPs may be neutralized through interdepartmental negotiations at the domestic level, before taking part in the international debate. CP representatives may also prevent strong attitudinal positions to increase the chance of reaching consensus.

Nonetheless, a closer look at the two sub-themes, discussed in Sect. 5.6.2, reveals that underneath the relatively neutral positions, substantially different human-nature relationship attitudes may be hidden. The analysis demonstrated that, in 2009, it has been possible to reach consensus on a set of general principles on Antarctic tourism, but it may well be that these principles do not exclude different attitudes of CPs. It is our impression that the differences in attitude do not cause problems in reaching consensus on the more general principles, while these different attitudes result in conflicting views in respect of the more specific management issues, such as the question of whether hotels in Antarctica should be prohibited. The general principles may leave enough space for different interpretations in accordance with different human-nature attitudes. For example, for some CPs the rejection of mastery requires a complete prohibition of hotels in Antarctica, while for others this would not be a necessary consequence. Germany and the Netherlands tend towards partnership, whereby the Antarctic environment, in principle, should not be affected by tourism. Other CPs, such as Japan, are more aligned with stewardship, whereby environmental care is taken but Antarctica may be permanently changed by humans.

The above discussion makes us aware of the fact that the ATCM focuses strongly on (the desirability of) norm setting. Many working papers discuss the need to adopt additional regulations on Antarctic tourism and often concrete proposals are attached in the form of a draft resolution or draft measure. However, norms are generally based on convictions about "values" (Schouten 2011) and – as far as the values of nature are concerned – it appears that such convictions on values are strongly determined by the human-nature relationships of the relevant actors (e.g., CPs). In other words, it is most likely that CPs with different human-nature relationships have different views on what the specific "values" of Antarctica are and how these values could best be protected, and, consequently, it is also most likely that these CPs have different opinions on what norms should be set in respect of specific tourism developments.

We see considerable merit in further studying this issue, for example, by conducting a more thorough and systematic review of policy documents as well as through qualitative interviews with representatives of CPs and expert organizations. Not with the ambition to "calibrate" the human-nature relationships as this appears quite unrealistic. However, more knowledge and consciousness of differences in human-nature attitudes would perhaps invite stakeholders to enter into the position of others to come to collective solutions based on a greater understanding of each other's convictions.

References

ASOC (2006a) Managing Antarctic tourism: a critical review of site-specific guidelines. Information Paper 65 presented by the Antarctic and Southern Ocean Coalition at the Antarctic Treaty Consultative Meeting XXIX, Edinburgh, UK

ASOC (2006b) Strategic issues posed by commercial tourism in the Antarctic treaty area. Information Paper 120 presented by the Antarctic and Southern Ocean Coalition at the Antarctic Treaty Consultative Meeting XXIX, Edinburgh, UK

ASOC (2008) A decade of Antarctic tourism: status, change, and actions needed. Information Paper 41 presented by the Antarctic and Southern Ocean Coalition at the Antarctic Treaty Consultative Meeting XXXI, Kyiv, Ukraine
AT (1959) Antarctic treaty. Available at: http://www.ats.aq/documents/ats/treaty_original.pdf
AT Secretariat (2011) Antarctic treaty Secretariat. Available at: www.ats.aq
ATCM (1994) Tourist guidelines. Recommendation XVIII-1 adopted at the XVIII Antarctic Treaty Consultative Meeting. Available at: http://www.ats.aq/documents/recatt/Att245_e.pdf
ATCM (2005) Final report of the XXVIII Antarctic treaty consultative meeting. Stockholm, Sweden. Available at: http://www.ats.aq/devAS/ats_meetings.aspx?lang=e
ATCM (2006) Final report of the XXIX Antarctic treaty consultative meeting, Edinburgh, UK. Available at: http://www.ats.aq/devAS/ats_meetings.aspx?lang=e
ATCM (2007a) Final report of the XXX Antarctic Treaty consultative meeting, New Delhi, India. Available at: http://www.ats.aq/devAS/ats_meetings.aspx?lang=e
ATCM (2007b) Long-term effects of tourism. Resolution 5 adopted at the XXX Antarctic treaty consultative meeting. Available at: http://www.ats.aq/devAS/info_measures_list_filtered.aspx?lang=e&cat=14
ATCM (2008) Final report of the XXXI Antarctic treaty consultative meeting. Kiev, Ukraine. Available at: http://www.ats.aq/devAS/ats_meetings.aspx?lang=e
ATCM (2009a) General principles of Antarctic tourism. Resolution 7 adopted at the XXXII Antarctic treaty consultative meeting. Available at: http://www.ats.aq/devAS/info_measures_list_filtered.aspx?lang=e&cat=14
ATCM (2009b) Final report of the XXXII Antarctic treaty consultative meeting. Baltimore, United States. Available at: http://www.ats.aq/devAS/ats_meetings.aspx?lang=e
ATCM (2010) Final report of the XXXIII Antarctic treaty consultative meeting. Punta del Este, Uruguay. Available at: http://www.ats.aq/devAS/ats_meetings.aspx?lang=e
ATCM (2011) Draft final report of the XXXIII Antarctic treaty consultative meeting. Buenos Aires, Argentina. The Final Report will become available at: http://www.ats.aq/devAS/ats_meetings.aspx?lang=e
Australia (2005) Protection of Antarctica's intrinsic values: policy on non-governmental activities. Working paper 38 presented at the Antarctic Treaty Consultative Meeting XXVIII. Stockholm, Sweden
Basberg BL (2005) Perspectives on the economic history of the Antarctic region. Discussion paper SAM 18–2005. Institutt for Samfunnsøkonomi, Bergen
Bastmeijer K (2003) The Antarctic environmental protocol and its domestic legal implementation. Kluwer Law International, The Hague
Bastmeijer K (2011) A long term strategy for Antarctic tourism: the key to decision making within the Antarctic Treaty System? In: Lück M, Maher P, Stewart E (eds) Polar tourism: human, environmental and governance dimensions. Cognizant Communication, Elmsford
Bastmeijer K, Roura R (2004) Regulating Antarctic tourism and the precautionary principle. Am J Int Law 98(4):763–781
Bastmeijer K, Lamers M, Harcha J (2008) Permanent land-based facilities for tourism in Antarctica: the need for regulation. Rev Eur Commun Int Environ Law (RECIEL) 17(1):84–99
Beeby C (1990) The convention on the regulation of Antarctic mineral resource activities and its future. In: Herr RA, Hall HR, Haward MG (eds) Antarctica's future: continuity or change? Australian Institute of International Affairs, Hobart, pp 47–60
Blay SKN (1992) Current developments, new trends in the protection of the Antarctic environment: the 1991 Madrid Protocol. Am J Int Law 86:377–399
CCAMLR (1980) Convention on the conservation of Antarctic marine living resources. Available at: http://www.ats.aq/documents/ats/ccamlr_e.pdf
CCAS (1972) Convention for the conservation of Antarctic seals. Available at: http://www.ats.aq/documents/recatt/Att076_e.pdf
Chen J (1995) Of agriculture's first disobedience and its fruit. Vanderbilt Law Rev 48(5):1262–1311

Chile (2009). The effect of marathons held on the Antarctic continent. Working Paper 54 presented at the Antarctic Treaty Consultative Meeting XXXII. Baltimore, United States

Cliteur P (2005) De filosofie van dierenrechten. In: Braeckman K, De Reuver B, Vervisch T (eds) Ethiek van DNA tot 9/11. Amsterdam University Press, Amsterdam, pp 135–159

CRAMRA (1988). Convention on the regulation of Antarctic mineral resource activities. Available at: http://sedac.ciesin.columbia.edu/entri/texts/acrc/cramra.txt.html

De Groot M, Drenthen M, De Groot WT (2011) Public visions on the human/nature relationship and their implications for environmental ethics. Environ Ethics 33(1)

Diamond J (1992) The third chimpanzee: the evolution and future of the human animal. Harper Collins Publishers, New York

Diamond J (2004) Collapse: how societies choose to fail or succeed. Viking, New York

Enzenbacher D (2007) Antarctic tourism policy-making: current challenges and future prospects. In: Triggs G, Riddell A (eds) Antarctica: legal and environmental challenges for the future. The British Institute of International and Comparative Law, London, pp 155–189

France (2006). Establishment of 'areas of special tourist interest'. Working Paper 18 presented at the Antarctic Treaty Consultative Meeting XXIX. Edinburgh, UK

Germany (2005). The admissibility of land-based tourism in Antarctica under international law. Information Paper 20 presented at the Antarctic Treaty Consultative Meeting XXVIII, Stockholm, Sweden

Haase D, Lamers M, Amelung B (2009) Self-regulation of tourism in Antarctica: exploring the conditions for success and failure. J Sustain Tour 17(4):411–430

Harrison P (1999) Subduing the earth: Genesis 1, early modern science, and the exploitation of nature. J Relig 79(1):86–109

Hayton RD (1960) The Antarctic settlement of 1959. Am J Int Law 54:349–371

Headland RK (2009) A chronology of Antarctic exploration: a synopsis of events and activities from the earliest times until the international polar year 2002–2009. Bernard Quaritch Ltd, London

Heap JA (1990) Antarctic sovereignty: a source of stress? In: Herr RA, Hall HR, Haward MG (eds) Antarctica's future. Tasmanian Government Printer, Hobart, pp 181–187

Holdgate MW (1987) Regulated development and conservation of Antarctic resources. In: Triggs GD (ed) The Antarctic treaty regime: law, environment and resources. Cambridge University Press, Cambridge, pp 128–142

IAATO (2004) Overview summarizing the terms of reference. Paper#12 presented at the Antarctic treaty meeting of experts. Tromso, Norway

IAATO (2011) Membership directory. Website of the International Association of Antarctica Tour Operators. http://apps.iaato.org/iaato/directory/list.faces. Accessed on 8 Sept 2011

Lamers M, Gelter H (2011) Diversification of Antarctic tourism: the case of a scuba diving expedition. Polar Rec 48:280–290 (FirstView)

Minteer BA, Manning RE (2005) An appraisal of the critique of anthropocentrism and three lesser known themes in Lynn White's "the historical roots of our ecologic crisis". Organ Environ 18:163–176

Netherlands and United Kingdom (2011) Antarctic tourism: towards a strategic and pro-active approach via an inventory of outstanding questions. Information Paper 21 presented at the Antarctic treaty consultative meeting XXXIII, Buenos Aires, Argentina

New Zealand (2005) Land-based tourism in Antarctica. Working Paper 12 presented at the Antarctic Treaty Consultative Meeting XXVIII. Stockholm, Sweden

Parkes G (2000) Precautionary fisheries management: the CCAMLR approach. Mar Policy 24:83–91

Pearson M, Stehberg R (2006) Nineteenth century sealing sites on Rugged Island, South Shetland Islands. Polar Rec 42:335–347

Protocol (1991) Protocol on environmental protection to the Antarctic treaty. Madrid, 4 October 1991. Retrieved 16 Feb 2011, from http://www.ats.aq/e/ats_protocol.htm

Schouten M (2011) Vormen van natuurbeheer en het behoud van biodiversiteit. Presentation at the D66 Conference 'Hoe verder met een duurzame natuur?' Arnhem, 1 oktober 2011

Schrijvers P (2008) De Natuur van de Dingen. Historische Uitgeverij, Amsterdam
Splettstoesser J (2000) IAATO's stewardship of the Antarctic environment: a history of tour operator's concern for a vulnerable part of the world. Int J Tour Res 2:47–55
Steenbakker P (1999) Benedictus de Spinoza (1632–1677), een overzicht. Filosofie 6:4–14
Stonehouse B, Crosbie K (1995) Tourist impacts and management in the Antarctic peninsula area. In: Hall CM, Johnston M (eds) Polar tourism. Wiley, New York
Thomas K (1983) Man and the natural world, changing attitudes in England (1500–1800). Allen Lane/Penguin Books Ltd, Harmondsworth
USA (2007) Approaches to tourism policy – next steps. Working Paper 6 presented at the Antarctic Treaty Consultative Meeting XXX, New Delhi, India
Van den Born R (2007) Thinking nature: everyday philosophy of nature in the Netherlands. PhD dissertation at the Radboud University Nijmegen. Available at: http://repository.ubn.ru.nl/bitstream/2066/55991/1/55991.pdf
Vicuña FO (1996) The regime of Antarctic marine living resources. In: Francioni F, Scovazzi T (eds) International law for Antarctica. Kluwer Law International, The Hague, pp 127–157
Wells S (2010) Pandora's seed: the unforeseen cost of civilization. Random House Publishers, New York
White L (1967) The historical roots of our ecologic crisis. Science 155:1203–1207
Zweers W (1995) Participeren aan de natuur; ontwerp voor een ecologisering van het wereldbeeld. Jan van Arkel, Amsterdam

Chapter 6
The Effectiveness of Environmental Impact Assessments on Visitor Activity in the Ross Sea Region of Antarctica

Tanya A. O'Neill, Megan R. Balks, and Jerónimo López-Martínez

Abstract The Madrid Protocol requires that before any kind of activities are conducted in Antarctica, the possible environmental impacts need to be assessed. Five case studies were investigated to assess the effectiveness of the Environmental Impact Assessment (EIA) system with respect to predicting and managing the effects of visitor activity in the Ross Sea region of Antarctica. Investigated EIAs included the decommissioning of the Greenpeace World Park Base, Cape Evans; Taylor Valley Visitor Zone, Taylor Valley; the decommissioning of the Vanda Station; occupation of the Loop Moraine campsite, both in the Wright Valley and McMurdo Dry Valleys; and the occupation and restoration of the ice-free storage area for the Cape Roberts Drilling Project, Cape Roberts.

The assessed sites were situated on a variety of landforms, soil parent materials, and local climates. At each site, the history of the site was known, and the time since last disturbance was well constrained. Assessment of the present state of the sites and accuracy of the EIA predictions compared with the actual impacts were carried out in the austral summers of 2008/2009 and 2009/2010 using visual soil assessment methods.

T.A. O'Neill (✉)
Landcare Research, Manaaki Whenua, Private Bag 3127, Hamilton, New Zealand

Department of Earth and Ocean Sciences, Waikato University, Private Bag 3105, Hamilton, New Zealand
e-mail: ONeillT@landcareresearch.co.nz

M.R. Balks, Ph.D.
Department of Earth and Ocean Sciences, Waikato University, Private Bag 3105, Hamilton, New Zealand
e-mail: megan.balks@waikato.ac.nz

J. López-Martínez, Ph.D.
Department Geologia y Geoquimica, Facultad de Ciencias, Universidad Autonoma de Madrid, 28049 Madrid, Spain
e-mail: jeronimo.lopez@uam.es

D.K. Müller et al. (eds.), *New Issues in Polar Tourism: Communities, Environments, Politics*, DOI 10.1007/978-94-007-5884-1_6,

At each of the five case studies, the standard of physical landscape recovery was higher than anticipated. Each case study demonstrated a high standard of EIA compliance, appropriate remediation of the site post-event, and follow-up monitoring where promised. Cleanup practices were of a high standard in each of the case studies investigated. The raking out of tracks, redistribution of disturbed stones, and careful backfilling of excavations all assisted in the rehabilitation of the formerly disturbed sites to near their predisturbed condition.

Although the EIA process may seem superficial, with a lot of form filling, the process is effective as it raises environmental awareness and motivates visitors to take responsibility for their "footprint" on the environment.

Keywords Environmental Impact Assessment • Ross Sea region • Soil surface recovery • Human impacts • Visual site assessment

6.1 Introduction

The persistence of cold even during summer, blizzards, severe winds, extreme aridity, and difficulties in accessibility are but a few of the factors that combine to curtail human activities in Antarctica. Yet we persist.

Antarctica was the last continent to be discovered. The historical explorations of Drake, Halley, and Cook in the 1600s and 1700s were followed by the first sealers and whalers in the Scotia Sea in the late eighteenth and nineteenth centuries and discovery of the South Shetlands Islands of the Antarctic Peninsula in the nineteenth century. It was not until the late nineteenth century where the impetus for scientific exploration of Antarctica was both encouraged and funded (Tin et al. 2009).

The upsurge in scientific activity of the early 1900s with the visits of Scott, Amundsen, and Shackleton culminated in the International Geophysical Year (IGY) in 1957–1958. By 2009, there were 40 year-round research stations in Antarctica (http://www.comnap.aq).

Tourism in the Antarctic region dates back to 1891, when the first tourists were passengers on resupply ships to the subantarctic islands (Codling 1995). Larger-scale commercial tourism began in 1957–1959 with Argentinean and Chilean naval transports, which accommodated tourists to help pay the costs of servicing national naval expeditions (Reich 1980). By 1966, Lars-Eric Lindblad began expedition cruising to the Antarctic Peninsula and initiated the use of zodiacs to ferry passengers to onshore sites. Antarctic-wide tourist numbers peaked in the 2007/2008 season with 46,265 tourists (approximately 32,000 landings). Due to expense, inaccessibility, and a lesser abundance of wildlife in the Ross Sea region, tourism remained at approximately 400 landing passengers per season from 2005/2006 to 2009/2010 (IAATO 2010).

The Protocol on Environmental Protection to the Antarctic Treaty, signed in Madrid in 1991 (The Madrid Protocol), designates Antarctica as a "natural reserve devoted to peace and science." The Madrid Protocol gives a comprehensive protection to the Antarctic environment and its ecosystems and maintains the value of the continent as an area for scientific research. The objective of this chapter is to use five case studies to illustrate the effectiveness of the current Environmental Impact Assessment system with respect to predicting and managing the effects of visitor activity in the Ross Sea region of Antarctica. Investigated Environmental Impact Assessments included:

- The decommissioning of the Greenpeace World Park Base, Cape Evans
- Taylor Valley Visitor Zone, Taylor Valley
- The decommissioning of the Vanda Station, Wright Valley
- Occupation of the Loop Moraine campsite, Wright Valley, McMurdo Dry Valleys
- The occupation and restoration of the ice-free storage area for the Cape Roberts Drilling Project, Cape Roberts

The similarities and differences between the Antarctic and Arctic Environmental Impact Assessment systems are briefly discussed.

6.2 Visitors to the Ross Sea Region

6.2.1 Tourism: Operators, Voyages, and Landing Sites

IAATO (the International Association of Antarctic Tour Operators) is a member organization founded in 1991 to advocate, promote, and practice safe and environmentally responsible private-sector travel to the Antarctic (http://www.iaato.org). IAATO member parties work within the Antarctic Treaty and Madrid Protocol, promoting Guidelines for Visitors to the Antarctic and coordinating member itineraries so that no more than 100 people are ashore in any one place at any one time. IAATO provides members with visitor guidelines, guidelines on wildlife watching, biosecurity, and decontamination guidelines for small boat operations.

Seaborne tourism in the Ross Sea region in the summer of 2009–2010 comprised three operators, who conducted five voyages, each involving helicopter and zodiac landings for a combined 397 passengers (IAATO 2010). A further eight voyages by three operators and an estimated 500 passengers were predicted to visit the area in the 2010/2011 season. In the last five years, 20 sites in the Ross Sea region have been visited by tourists; the 12 most visited sites, and their attractions, are summarized (Table 6.1).

6.2.2 National Program Visitors

Four of the twelve sites most frequently visited by tourists (Cape Royds, Cape Adare, Cape Evans, and Hut Point Peninsula) are also regularly visited by New

Table 6.1 The 12 most visited sites in the Ross Sea region

	Site and attraction	Mean annual visitors numbers[a]
1	*Ice landings*: includes landings on the Drygalski Ice Tongue, B15K, Ice walks near Cape Hallett, Cape Royds, Beaufort Island, and the Ross Ice Shelf	367
2	*Cape Royds*: Shackleton's Nimrod Hut. Adélie penguins	318
3	*Cape Adare*: Borchgrevink's and Scott's Northern Party Huts. Adélie penguins	293
4	*Cape Evans*: Scott's Terra Nova Hut	291
5	*Franklin Island*: Adélie penguins	215
6	*Terra Nova Bay*: Mario Zuchelli Station Tour and Vantage point for Inexpressible Island	192
7	*Inexpressible Island*: His toric site where Scott's Northern Party from the Terra Nova Expedition wintered in a snow cave. Adélie penguins	154
8	*Hut Point Peninsula*: Scott's Discovery Hut, McMurdo Station, Scott Base and Observation Hill	154
9	*Cape Washington*: Emperor penguin colony	154
10	*Coulman Island*: Emperor penguins and historic site	115
11	*Cape Hallett*: Adélie penguins	110
12	*Canada Glacler (Taylor Valley Visitor Zone)*: Dry Valleys experience: Canada Glacier, mummified seals, ventifacts	109

Data from: Antarctica New Zealand (2009)
[a]Mean annual tourist visitors between 2005/2006 and 2008/2009

Zealand and US National Program personnel and base staff (Table 6.2). Tourist visitors comprise the greater proportion of visitors to Cape Royds and Cape Adare each year, whereas Cape Evans is visited by a large number of US National Program personnel, who also comprise the largest proportion of visitors to Scott's Discovery Hut on the Hut Point Peninsula. McMurdo Station's close proximity to Discovery Hut makes Discovery Hut a "must-see" attraction among base staff and other US National Program personnel on their off-days.

6.3 Environmental Impact Assessment

Annex I of the Madrid Protocol requires that before any kind of activities are conducted, the possible environmental impacts need to be assessed. The term Environmental Impact Assessment (EIA) was introduced in Article 8 of the protocol and is a process whereby persons responsible for an activity predict the likely environmental impacts and assess their significance (Committee for Environmental Protection 2005). The EIA process is at one of three levels, depending on the nature and scale of the activity (from http://www.antarcticanewzealand.govt.nz):

Table 6.2 The total number of visitors, US National Program, New Zealand National Program, and tourists, to Cape Royds, Cape Adare, Cape Evans, and the Hut Peninsula, in the Ross Sea region

	Annual visitor ratios to most frequented sites				Mean annual
Site and attraction	2005/2006	2006/2007	2007/2008	2008/2009	Visitor numbers[a]
Cape Royds: Hut. Adélie penguins					
US National Program	125	215	162	98	150
New Zealand National Program	174	133	167	47	130
Tourists	369	454	171	282	319
Total	**668**	**802**	**500**	**427**	**599**
Cape Adare: Huts. Adélie penguins					
US National Program	0	0	0	0	0
New Zealand National Program	2	0	0	0	1
Tourists	538	19	119	465	285
Total	**540**	**19**	**118**	**465**	**286**
Cape Evans: Hut					
US National Program	153	557	616	257	396
New Zealand National Program	167	131	192	73	141
Tourists	278	359	171	272	270
Total	**598**	**1, 047**	**979**	**602**	**807**
Hut Point Peninsula: Hut. Stations					
US National Program	639	619	627	583	617
New Zealand National Program	105	138	89	38	93
Tourists	423	19	133	250	206
Total	**1, 167**	**776**	**849**	**871**	**916**

Data from*:* Antarctica New Zealand (2011b)
[a]Mean annual ratios of National Program and tourist visitors between 2005/2006 and 2008/2009

1. *Preliminary environmental evaluations (PEEs)* are processed at the national level and required where impacts are likely to be less than minor or transitory.
2. *Initial environmental evaluations (IEEs)* are notified to the Antarctic Treaty Parties and required where impacts are likely to be minor or transitory.
3. *Comprehensive environmental evaluations (CEEs)* are considered by the Antarctic Treaty Parties and required where impacts are likely to be more than minor or transitory.

Project leaders must assemble and analyze information on the potential environmental effects a proposal may have and how they can be best prevented or mitigated. The EIA should be undertaken before major decisions are made, ideally, while feasible alternatives and options to the proposal can still be actioned. The people responsible for an EIA need to consult as widely as is reasonably necessary and possible in order that the best available information and professional advice contribute to the outcome (Committee for Environmental Protection 2005).

Where a CEE is necessary, such as the multinational Cape Roberts Drilling Project (1997–1999), the draft CEE was made publically available and considered by the Committee for Environmental Protection, which then advises the Antarctic Treaty Consultative Meeting. The Antarctic Treaty Consultative Meeting is a mechanism established by the Antarctic Treaty which over the last decade meets annually for the purpose of exchanging information, consulting and formulating recommendations on many matters including current CEE applications, and thus ensuring the principles and objectives of the Treaty and Madrid Protocol are fulfilled. During the Antarctic Treaty Consultative Meeting, the CEEs for the proposed activities are discussed, comments are provided by other Treaty signatory representatives, and a decision is made as to whether and how an activity is conducted.

The main purpose of the EIA is to facilitate the systematic consideration of environmental issues as part of development and decision-making (Abaza et al. 2004).

6.4 New Zealand's Visitor Site Assessment Scheme (VISTA)

EIA is based on the premise that activity-related environmental impacts can be predicted. But in reality, our current understanding of cumulative impacts and cause-effect relationships in Antarctica remains limited. Because of this, a precautionary approach should be taken where the likely impacts of the proposal are unknown or uncertain.

Antarctica New Zealand administers a visitor site assessment scheme (VISTA), which aims to support the EIA process and address the shortfall of information on the cumulative impacts of visitor activity through a site monitoring program. Representatives use a series of booklets, maps, and photos to help orientate themselves at a site and find waypoints for photo and ground disturbance monitoring sites. Annual replication of fixed photo points allows changes such as site recovery, or cumulative disturbance, over time to be monitored. Ground disturbance or "terrestrial impact visual assessments" are also carried out at all landings on dry ground. The visual site assessment method of Campbell et al. (1993) was used in assessing surface disturbance in 5-m × 5-m plots in areas where visitors walk. Where possible, the assessment is carried out before the visitors land and repeated after their visit has been completed. Information on wildlife, vegetation, evidence of previous ground tracking, and other observations are also collected to give an overview of the environmental sensitivities of the site. Currently 20 sites in the Ross Sea region are part of the VISTA monitoring program (Antarctica New Zealand 2009), including the Taylor Valley Visitor Zone, one of the sites used as a case study in this chapter.

6.5 Examples of New Zealand Managed Environmental Impact Assessments

Under New Zealand's Antarctica (Environmental Protection) Act 1994 (which implements Annex I (Madrid Protocol) provisions), all activities organized in New Zealand or proceeding to Antarctica from New Zealand must undertake an EIA. The evaluation must be approved by the Antarctic Policy Unit of the Ministry of Foreign Affairs and Trade (MFAT) before the activity can commence.

A Crown Entity established in 1996 under the New Zealand Antarctic Institute Act (1996), known as Antarctica New Zealand, is responsible for developing, managing, and executing New Zealand government activities in Antarctica and the Southern Ocean. Antarctica New Zealand is required to develop, manage, and execute New Zealand's activities in the Ross Sea region of Antarctica, conserve the intrinsic value of Antarctica by understanding and managing human impacts on the Antarctic environment, and raise public awareness of the international significance of the continent. The environmental impacts of an "event" managed by Antarctica New Zealand should never outweigh the likely benefits of the scientific research and its outcomes (Antarctica New Zealand, Statement of Intent 2011a).

The effectiveness of the EIA process was investigated over the 2008/2009 and 2009/2010 field seasons at five Ross Sea region sites of human occupation and activity (Table 6.3, Fig. 6.1).

The assessed sites were situated on a variety of landforms, soil parent materials, and climates (Table 6.3). At each site, the history of the site was known, the time since last disturbance was well constrained with sufficient background information into the EIA predicted impacts, and, in some cases, photo documentation was available. Assessment of the present state of the sites and accuracy of the EIA predictions compared with the actual impacts were carried out in the summers of 2008/2009 and 2009/2010.

6.6 Site of the Greenpeace World Park Base

The ice-free area of Cape Evans, Ross Island, is a site of historical value, with Scott's 1910 Terra Nova Hut and associated artifacts as well as other monuments such as the Cross erected by Scott's expedition on Wind Vane Hill. In the summer of 1986/1987, Greenpeace established a base 400 m along the beach from Scott's Terra Nova Hut on a raised beach terrace about 5 m above sea level between low rock outcrops and the present day beach (Fig. 6.2a). The site had been previously occupied on a number of occasions since Scott's arrival in the early 1900s. The last occupants had been members from Robert Swan's "Footsteps of Scott" British expedition to South Pole in 1984–1986, where the crew set up camp at the site

Table 6.3 General location, description, history, landform, climate, and estimated annual visitor numbers at the case study sites

Site description and history	Type of EIA	Time since last disturbance	Post-EIA monitoring	Estimated visitor #s	Landform	Parent material	Ross Sea region climate zone[a]
Former Greenpeace World Park Base (77° 38′ 05.2″ S, 166° 25′ 20.6″ E): a 4-person base established on a pre-disturbed site in the summer of 1986/1987. The station was decommissioned, removed and the former site remediated in the summer of 1991/1992.	IEE	13 years	1991/1992 1992/1993 1995/1996	Site currently occupied by Antarctic Heritage Trust. Up to 8 people in the summer	Active beach	Scoriaceous basalt and locally derived till, alluvium, colluvium and beach gravels	Moist coastal mountain
Taylor Valley Visitor Zone (77° 37′ 35.3″ S, 163° 03′18.6″ E): Designated as a Visitor Zone in the McMurdo Dry Valleys; spectacular landscape for visitors to see the Canada Glacier ter minus, mummitied seals, ventitacts and patterned ground.	IEE	10 months	Yes VISTA	224 (2006/2007) 117 (2007/2008) 93 (2008/2009)	Glacial moraine	Mixed till	Central mountain
Loop Moraine field site (77° 29′ 10.8″ S, 162° 21′ 50.6″ E): Campsite for New Zealand Programme K123 during summer 2004/2005; occupied by 4 people for 1 month. Campsite, and walking track formed from the campsite to toilet facilities, and stream-side experiment.	PEE	5 years	No	Not in last 5 years	Glacial moraine-alluvial fan	Mixed till and alluvial fan deposits	Central mountain

Former Vanda Station site (77° 31′ 41.5′′ S, 161° 40′ 18.6′′ E): established in the summer of 1968/1969 at Lake Vanda, Wright Valley. The station was occupied predominately during summers until 1990/1991, and decommissioned in 1994/1995 due to rising lake levels at the near by Lake Vanda ([a]lake level and contaminants monitoring 1994–2002)	IEE	17 years	Yes[a]	<10 annually	Low ridge on the valley floor	Surface lag layer of mixed till over granodiorite blocks, bedrock, erosion-resistant lamprophyre dykes	Central mountain
Former Cape Roberts Project storage area (77° 02′ 07.3′′ S, 163° 10′ 43.020′′ E): Ice-free storage site for the Cape Roberts Drilling Project from 1997 through 1999. Site remediation began in 1999/2000 and the disturbed site was manually raked in 2000/2001.	CEE	10 years	Yes	<5 annually	Active beach-bedrock outcrop	Beach deposited gravels and granodiorite bedrock	Moist coastal mountain

[a]Climate zones are those of Campbell and Claridge (1987)

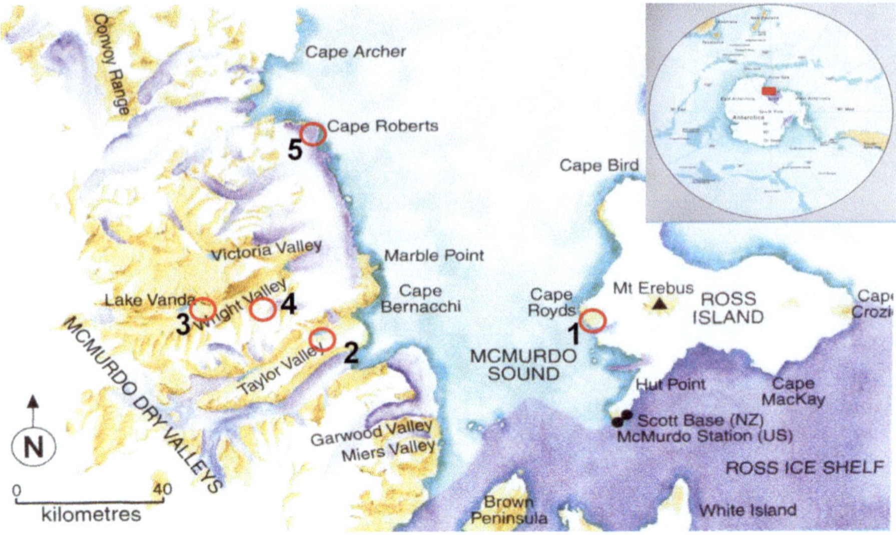

Fig. 6.1 Location of Ross Sea region case study sites: (*1*) Greenpeace Base, Cape Evans; (*2*) Taylor Valley Visitor Zone; (*3*) Vanda Station, Wright Valley; (*4*) Loop Moraine campsite, Wright Valley; (*5*) Cape Roberts storage area, Cape Roberts (not drawn to scale)

for number of days before recreating Scott's last march to the Pole (Mear and Swan 1987). The Greenpeace four-person winter-over base consisted of a main building of prefabricated units constructed in 1987, and then from 1988 to 1990, several structures were added to the base, forming an L-shaped main building, fuel rack, storage shed for food, SatCom tower, wind generator, and radio antennas (Greenpeace 1991a).

6.6.1 *Environmental Impact Assessments and Restoration Efforts*

The IEE for the commissioning of the base stated that potential impacts during the construction phase could include fuel spills and stress on the immediate surrounds, including disturbance to skuas and the ground, as well as dust and helicopter noise pollution (Greenpeace 1986). During occupation and overwintering of the base, the EIA stated that the "presence of four people in a limited area will put some limited stress on the immediate surroundings" but that the small party would have little impact on the soils, that there was "no nearby vegetation, and that no vehicles were planned to be used inland." Greenpeace concluded that the "foreseeable environmental impacts of the proposed activities of base construction are negligible, and an Environmental Impact Statement [precursor to a CEE] is not needed." Monitoring during site occupation and at the conclusion and decommissioning phases of the program was promised.

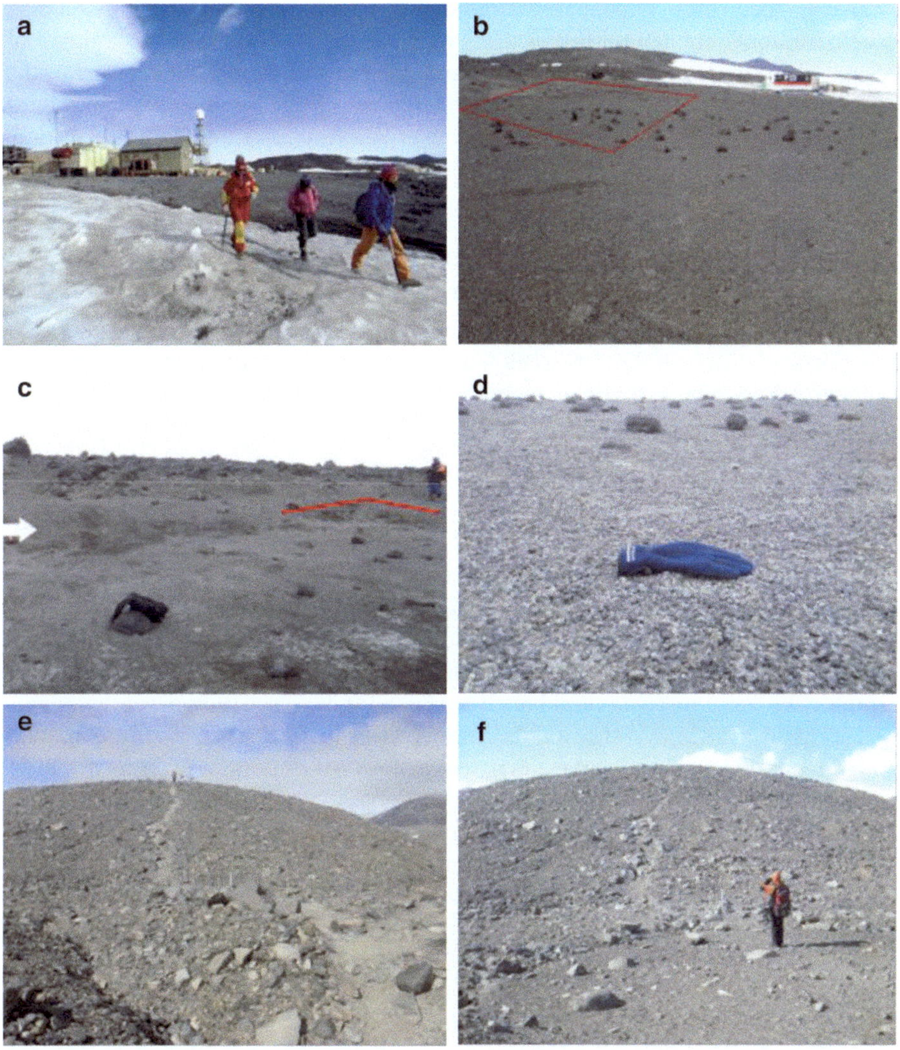

Fig. 6.2 Sites of Greenpeace World Park Base at Cape Evans and the Taylor Valley Visitor Zone. (**a**) Greenpeace huts and fuel storage racks in January 1992; (**b**) *red rectangle* marking the location of the former Greenpeace base; (**c**) *red marks* the trace of former corner of main Greenpeace building and unnatural gravel mounds; (**d**) "unnaturally" placed boulders on top of the raised beach surface, with recovered surface in foreground (*glove* for scale); (**e**) Taylor Valley Visitor Zone steep track to lookout, 06/07 season, post-96 visitors; (**f**) Taylor Valley Visitor Zone steep track to lookout, 09/10 season, before first tourist group of the season

An IEE was prepared for decommissioning of the Greenpeace base, and removal and remediation of the site took place in the 1991/1992 summer. The IEE indicated that "the proposed base removal will not have more than a minor and transitory effect on any component of the environment, and a CEE is not needed" (Greenpeace 1991a) and that ongoing monitoring of the site would be continued beyond the 1991/1992 decommissioning season.

During occupation, Greenpeace estimated that a maximum of 30 m^3 of soil may have been contaminated with fuel. Contaminated soil was removed to a depth of 20 cm in the most contaminated sites and returned to New Zealand during the site remediation process to ensure all practicable mitigative pathways had been followed to prevent contamination via subsurface flow into the Ross Sea (Roura 2004). In total, it is estimated that 5 tonnes of contaminated sediments were removed (Greenpeace 1992, 1994b). All excavations were backfilled and leveled to prevent permafrost degradation. Most of the sediments used as backfill were extracted from the present day beach, where the potential for permafrost degradation was comparatively minor (Roura 2004). Environmental monitoring to evaluate the likelihood of fuel migration and to assess backfill stability was carried out by Greenpeace in 1991/1992, 1992/1993, and 1995/1996 with the logistic support of Antarctica New Zealand (Greenpeace 1991a, b, 1992, 1993, 1994a, b, 1997).

6.6.2 *Observations*

Photos and sketches were used as a means of locating the site of the former Greenpeace base, and the geomorphological structures of the area were identified. Soil was excavated to the depth of ice-cemented permafrost (approximately 40 cm below the surface in January 2009) around the vicinity of the former fuel rack, and a qualitative "sniff test" was conducted to check for evidence of hydrocarbons.

The main visible evidence of the former Greenpeace base was the unnatural gravel mounds associated with either the removal of persistent snow drifts from the landward side of the building or site flattening to accommodate the base (Fig. 6.2b, c). Unnaturally placed boulders on top of the raised beach surface were noted (Fig. 6.2d) as well as an occasional nail or small (<10 cm long) piece of wood (though the litter may be attributable to the Scott era). On an undisturbed surface, the rocks would be embedded into the pavement, whereas rocks sitting clear on top of the surface appear to have been placed there in an attempt to restore the surface. A trace of a corner of the main building was still visible, though not obvious, where the area was leveled and Greenpeace cut into the ridge (Fig. 6.2c).

Overall, there was little remaining evidence of the former base. It took knowledge of the location of past activities and careful observation to identify any evidence of disturbance. A lag gravel deposit had reformed over the entire site. There was no evidence of structures such as the drainage trench, SatCom tower, or food storage hut. The raised beach surface showed no evidence of permafrost degradation in backfilled areas, no visual expressions of fuel spills, and no remnants of walking tracks or drains, or any obvious differences in texture or surface color when compared with nearby undisturbed beach surfaces.

The beach worked loose gravel at the site is readily resorted by surface wind, water, freeze-thaw, and snow melt and runoff processes, making it an active and readily recoverable surface. The site was temporarily reoccupied in 2011 by the Antarctic Heritage Trust while they carry out restoration of Scott's Terra Nova Hut.

6.7 Tourism at the Taylor Valley Visitor Zone

The Taylor Valley Visitor Zone (TVVZ) is located adjacent to the Canada Glacier in the Taylor Valley. Designated as a "tourism zone" in 1995, the TVVZ is currently the only area in the McMurdo Dry Valleys where tourists are allowed to visit. It is an attractive site which provides visitors with a spectacular experience of a typical "Dry Valley" site. The visitor zone was selected as a site where "safe and easy access and movement within the area can be reasonably assured with minimal impact to science activities or the environment" (New Zealand, the United Kingdom, the United States 2006). Access is by helicopter only. The tourism zone is approximately 350 m × 300 m, dominated by a low (approximately 30 m high) hill, and is covered in gravely rocky till. There are established walking tracks and small guided groups walk around the site, passing by a proglacial lake and the terminus of the Canada Glacier, up a steep track to a lookout point, and also past mummified seals. A total of 434 people have landed between the summers of 2006/2007 and 2008/2009. Average landing times are about 9 h (entire party) (Antarctica New Zealand 2009). Research program activity also occurs in the vicinity of the site, with substantial United States Antarctic Program camps at nearby Lake Fryxell and Lake Hoare, so total visitor numbers are difficult to quantify.

6.7.1 Environmental Impact Assessment

Inspection of a recent IEE from a tourist operator working in the Ross Sea revealed the sections relating to the assessment of impacts to surfaces traversed and possibilities of cumulative impacts as generic, stating the activities proposed in Antarctica "are likely to have less than minor or transitory impact on the Antarctic environment." Details regarding specific sites to be visited and precautions required at each site were not included. Perhaps this is typical of tourist operator IEE and was confirmed by an Antarctic Policy Unit compliance officer from the New Zealand Ministry of Foreign Affairs and Trade (Ane Hunkin, pers. comm. June 21, 2011).

6.7.2 Observations

A visual site assessment and Antarctic New Zealand-administered visitor site assessment (VISTA) scheme photo-replication revealed that the physical impacts of the tracks within the TVVZ persist between seasons (Fig. 6.2e, f), although distinct footprints were not evident prior to tourists arriving in December 2009. An overall contrast in desert pavement color between the track and adjacent undisturbed mixed till pavement was evident, and due to the concentration of larger clasts being pushed to the outer margin, the formed walking track was clearly evident. However, VISTA photo-replication showed that the color contrast between track and adjacent control was much less pronounced after the site had a winter to recover, implying that some desert pavement recovery does occur between tourist seasons. It was also observed that the tracks follow the patterned ground cracks were practicable. The patterned ground cracks are the most active and readily recoverable areas in the landscape and the most favorable place to concentrate visitor flow.

6.8 Site of the Former Vanda Station

Vanda Station was established adjacent to Lake Vanda in the 1968/1969 summer to support research in the Wright Valley (Fig. 6.3a). Human occupation left the former Vanda Station site disturbed by trampling, vehicle movement, excavations, occasional oil spills, and accumulation of debris (Webster et al. 2003). Urine and gray water was strained and dumped in an adjacent gully until 1993. With rising lake levels, the potential for leaching of contaminated ground into the lake system was a concern (Sheppard et al. 1993), so decommissioning of the station took place from the summers of 1992/1993 to 1994/1995 (Hayward et al. 1994).

6.8.1 Environmental Impact Assessment and Restoration Efforts

An IEE was submitted to carry out the decommissioning and remediation of the site in March 1994. The IEE stated that all physical structures would be removed; all contaminated soil with the potential to adversely impact on lake waters would also be removed. And in addition, a monitoring program would be undertaken to ensure that any changes in lake levels could be identified (Hayward et al. 1994). Little information was included in the IEE regarding the impacts from the removal activities. All that was mentioned was that the decommissioning activities would "have the minor and transitory impacts of generating dust and noise" and that "all solid and liquid wastes generated during this period will be removed . . . " (Hayward et al. 1994).

Over the summer of 1994/1995, a team of 4–6 people were stationed at the site. Removal of the station, comprising eight buildings able to accommodate 14 people, took over 180 person days and 70 helicopter hours to complete. Buildings and a large amount of contaminated soil and painted rocks were removed (Webster et al. 2003). Effort was also made to restore land surfaces to a "natural" condition including pathways and tracks (Hayward et al. 1994). The IEE indicated that a significant environmental concern arising from the removal may activate soil contamination in the immediate vicinity of the former buildings (Hayward et al. 1994). The risk of leaving the structures and contaminated soil in place, however, with rising lake levels causing major contamination into the lake waters, was considered to outweigh the risk (Sheppard et al. 1993).

Fig. 6.3 Site of the former Vanda Station, the Loop Moraine campsite, and Cape Roberts ice-free storage area. (**a**) Vanda Station in January 1992 immediately prior to removal; (**b**) the site of the main Vanda mess hut. The three people are marking the corners of the buildings; (**c**) Loop Moraine campsite, December 2004. Photo: Fiona Shanhun; (**d**) Loop Moraine campsite, 5 years later, December 2009; (**e**) Loop Moraine track to toilet, track margin visible (*white arrows*), after 5-year recovery, December 2009; (**f**) Cape Roberts storage area, c. 1997. Photo: Rebecca Roper-Gee; (**g**) Cape Roberts remediation January 2001. Photo: Rebecca Roper-Gee

Following decommissioning of the site, Antarctica New Zealand undertook an annual program of water and algae sampling from 1994/1995 until 2001/2002. Samples were taken from Vanda Bay (immediately adjacent to the station site) and a control site to ensure any changes in lake water chemistry could be identified. Analysis of data, including dissolved nutrients, chlorophyll, particulate nutrients, algal species composition, and heavy metals, revealed that there was "no evidence to date to suggest contamination has occurred in Vanda Bay" (Sutherland 2002).

6.8.2 *Observations*

The hardy nature of the bedrock and the thorough cleanup made it difficult to locate the site of the former central court and mess hut (Fig. 6.3a, b), even though a member of our party had stayed at the site on a number of occasions. There were no visible footprints, all building materials had been removed, displaced rocks had been redistributed, and efforts had been made to redistribute boulders concentrated along vehicle and walking track margins. It was noted that black lamprophyre dyke rocks, which outcrop in long narrow stripes on the surface, were redistributed somewhat unnaturally among the lighter granitic slabs and bedrock.

6.9 Loop Moraine Campsite and Walking Tracks

The New Zealand Antarctic Program K123 Loop Moraine campsite and associated walking tracks were occupied for 1 month by four people during the summer of 2004/2005 (Fig. 6.3c). The camp consisted of three tents and a nearby site where toilet buckets were placed. Also field equipment was stacked adjacent to the site of three helicopter landings.

6.9.1 *Environmental Impact Assessment and Restoration Efforts*

The event IEE indicated that there would be disturbances to ice-free areas but that "any soil sampling pits would be backfilled, and desert pavement restored as much as possible, [and that] any stones moved would be replaced." The IEE detailed that impacts from the visit would be minimal as the environments visited had "active surfaces from which disturbances are soon obliterated" and stated that their total impact on the environment is likely to be "less than minor and transitory" (Preliminary Environmental Evaluation K123 2004).

At the end of January 2005, all tents and equipment were removed, and any obviously disturbed stones were returned to their original orientation. At the time of departure from the site, footprints were widely evident around the area where the tents had been, and distinct walking tracks had formed between the campsite and the toilet facilities and the campsite and streamside experiment (Fig. 6.3e)

6.9.2 Observations

At the Loop Moraine, widespread trampling around campsites was undetectable within 5 years (Fig. 6.3c, d), yet evidence of walking tracks remained in the landscape (Fig. 6.3e), even though the desert pavement had recovered. The larger clasts which concentrated along the margins of confined tracks remain visible in the landscape (Fig. 6.3e).

6.10 Cape Roberts Ice-Free Storage Area

Cape Roberts is a small ice-free coastal peninsula situated on the southern side of Granite Harbor on the Antarctic Mainland. Currently, the facility zone includes an area of flat ground where two huts, a toilet shelter, and a fuel depot are located. The buildings were erected in November 1984, and the site used for winter storage of vehicles, drilling, and camp equipment during the Cape Roberts Drilling Project, a multinational marine sediment drilling program that operated from 1995 to 2000 (Fig. 6.3f). Site remediation occurred over the summers of 1999/2000 and 2000/2001. The huts were left in place to support ongoing tide gauge measurements (Fig. 6.3g). Since site remediation, the numbers of visitors to Cape Roberts have ranged between 1 (summer of 2003/2004) and 20 visitors (summer of 2009/2010) (Antarctica New Zealand 2011b).

6.10.1 Environment Impact Assessment and Restoration Efforts

A baseline survey of the Cape Roberts storage area was carried out over the summer of 1992/1993 to collect environmental data which would enable the impact of the project to be assessed at its conclusion (Campbell and Keys 1993). General soil and vegetation patterns were determined, and representative plots established in the areas deemed likely to be subject to the highest impacts. A CEE was required for the Cape Roberts Drilling Project. The CEE predicted that the main impacts at Cape Roberts would be caused by vehicles and storage of materials and human trampling in an area previously disturbed by human activities.

Likely impacts would include surface disturbance by compaction as well as the risk of contamination from fuel spills. Helicopter movement was also predicted to impact the skua populations in the vicinity. The CEE concluded that there would be cumulative impact resulting from the proposed activity; however, it would be "small and insignificant in relation to its advantages" (Keys et al. 1994). A decommissioning, spill monitoring, and a cleanup phase took place over the summers of 1999–2001, and rakes were used to redistribute rocky soil disturbed by vehicle tracks and other activities. Full restoration of the storage area to pre-project condition was attempted (Fig. 6.3g).

6.10.2 Observations

The site was visited in January 2010, 10 years after remediation took place. A visual site assessment was repeated on the representative plots first surveyed pre-project in 1992/1993, and evidence of former site occupation was undetectable. It was evident that the use of manual raking in site remediation had been effective.

6.11 Discussion

The five case studies showed that overall, the standard of landscape recovery was higher than anticipated. The decommissioning and site remediation of the Greenpeace World Park Base at Cape Evans, and Vanda Station, in the McMurdo Dry Valleys, was effective. There was very little evidence of either base and no evidence of former structures or vehicle tracks. A considerable volume of contaminated soil was removed at both the Greenpeace World Park Base and the Vanda Station during site remediation, and we found no evidence of permafrost degradation or humps or hollows indicating backfilled areas. The faint trace of the corner of the Greenpeace base would not be noticeable to the casual observer.

It can be assumed that generally, the cleanup at Greenpeace World Park Base would have required less effort than that of Vanda Station as the EIA system was in place and precautionary measures, for example, to prevent oil spills, would have been put in place from the commissioning of the station in 1987. Life at Vanda Station, on the other hand, was different in the late 1960s, with few measures in place to mitigate the environmental concerns we have today.

The Taylor Valley Visitor Zone is the only zone in the McMurdo Dry Valleys that is visited by tourists. Tourist operators in the Ross Sea region are careful to follow the guidelines set for the visitor zone, and preexisting walking tracks are used. Use of patterned ground cracks as the preferred route for walking tracks was effective as the active movement of materials in the patterned ground cracks assists surface recovery. Previous work in the McMurdo Dry Valleys, by Campbell et al. (1998), has shown that once a track has formed (within 20 or so passes), the cumulative impacts of larger numbers of people following the same track (20 passes versus 200 passes versus 2,000 passes) are minimal. The use of a single confined track will preserve the existing undisturbed desert pavement adjacent to current tracks, preserving the natural integrity of the landscape. The cumulative impacts of repetitive visitation are thus minimized by keeping people to one clearly defined walking track. We found VISTA an effective tool to assess cumulative impacts and recovery between seasons at the Taylor Valley Visitor Zone. VISTA photo-replication showed that the physical impacts of the tracks within the visitor zone persist between seasons. However, distinct footprints were not evident.

The Loop Moraine campsite is an excellent example of the rate of recovery in areas of the McMurdo Dry Valleys. Previous studies have shown that most of the ice-free areas in the Ross Sea region of Antarctica are readily disturbed by human activities and slow to recover in the prevailing cold and arid conditions (Campbell and Claridge 1987; Campbell et al. 1993, 1994, 1998; Claridge et al. 1995). In some cases, visual impacts can persist in the landscape for a very long time (Campbell and Claridge 1987; Campbell et al. 1993, 1998). The Loop Moraine campsite challenges this long-standing observation as we observed the widespread trampling around the campsite was undetectable within 5 years. The scale of the physical impact will naturally have some bearing on recovery time, such as disturbance from bulldozer blades compared with pedestrian foot traffic; the former persists in the Antarctic landscape for decades (Campbell and Claridge 1987; Campbell et al. 1993, 1998). The Loop Moraine site also illustrated the differences in recovery rates of confined tracks and randomly trampled sites. Evidence of the campsite itself was undetectable; however, well-used tracks to and from the campsite were visible 5 years on, and in hindsight, the group may have chosen to either rake out formed tracks or walk to their experimental sites and the toilet facilities in a random and widespread fashion, avoiding the formation of confined walking tracks.

Finally, the CEE for the Cape Roberts Drilling Project was accurate, and as predicted, surface compaction and formation of vehicle and walking tracks did result from site activities. The CEE also stated that site restoration would take place, and 10 years on, the impacts of the former occupation were not visible.

6.11.1 Cleanup Practices

Observations at the site of the Cape Roberts storage area show that the remediation efforts, in particular, the use of surface raking, were effective. A rake was used to redistribute surface stones which had concentrated in areas of vehicle activity and foot traffic as well as to loosen compacted areas. The technique led to recovery of the desert pavement within 9 years.

The redistribution of slabs of granodiorite and lamprophyre dyke rocks at the Vanda Station site, and volcanic boulders at the former Greenpeace base site, were effective forms of remediation, although repositioning did look unnatural in some cases. Redistribution of disturbed rocks needs to be undertaken with an understanding of their natural position in the environment, and rocks embedded back into the pavement surface for a natural appearance. It is also important to note that there was no relationship between the amount of time the site was occupied and the effectiveness of site rehabilitation, as evidenced when we compare the 5-year occupation of the Greenpeace base site with the 25-year occupation of the Vanda Station site and 1 month at the Loop Moraine site. Former occupation was almost undetectable at all sites, and in all cases, recovery was a function of the effort of the personnel doing the cleanup and the active wind- and water-driven surface processes.

6.11.2 Environmental Impact Assessment Effectiveness

Investigation into five sites in the Ross Sea region has demonstrated that overall appropriate site restoration and post-event monitoring are being conducted with an excellent standard of outcomes. On the exterior, the EIA process may seem superficial with a lot of generic form filling. The scheme's greatest asset is that it raises awareness by increasing accountability and through motivating members of events to take responsibility for their activities.

EIA follow-up investigations such as this are often a neglected element of the EIA process. Effort should be made to systematically carry out evaluations on a regular basis to ensure the process remains working effectively.

6.11.3 Comparison of Antarctic and Arctic Environmental Impact Assessment Systems

The Antarctic and Arctic show similarities in some respects; both regions are under growing environmental stress and have fragile elements susceptible to damage from outside sources, but in terms of environmental protection, sustainability, and international environmental law, the poles differ substantially. The Antarctic, whose environmental legal regime is governed by the Antarctic Treaty System (1959), has been called the model in international law for a purely environmentally focused management regime (Nowlan 2001). A "freeze" on sovereignty claims has resulted in a single comprehensive environmental protection regime, whose international environmental regulations have then been incorporated into the national legal systems of the Antarctic Treaty Consultative Parties. Furthermore, the development of the legal system occurred under the absence of an indigenous human population, permanent population, or industrial and resource conflicts, thereby making execution of its principles easier in some respects. The Arctic, on the other hand, is a region dominated by the existing national legal systems of the eight

Arctic states. Complicating environmental management are the approximate 10 million people that live in what is defined the Arctic region of whom 1.5 million are of indigenous origin, most of whom require economic activity for survival. Exploitation of natural resources, particularly mining, is prohibited by the Antarctic Treaty, whereas mineral exploration and development are well advanced in Canada, the United States, Norway, and Russia, and mining activities are subject to the domestic legal regulatory regime of each Arctic state. The inhabitants of the Arctic therefore need to live under a legal regime that permits them to thrive, while at the same time protecting the fragile environment. In the Arctic, domestic law governs environmental assessment, though the Arctic states are encouraged to follow the voluntary EIA Guidelines (Nowlan 2001). EIAs are undertaken for most large-scale activities and applied to activities associated with the exploitation of natural resources, public use, military activities, and the development of infrastructure that may cause significant environmental impacts (Nowlan 2001). The EIA assessment, scope, considering alternatives, collection of baseline data, predicting cumulative impacts, and monitoring during and after the activity are essentially the same, however, in the Arctic public participation and consultation between indigenous peoples, groups, organizations, and communities, which have an interest in, or could be affected by, the proposed action is a requirement. Inclusion of public consultation into the EIA process in the Arctic means all those affected by the project are able to provide input into the planning, ensure openness, establish trust among all parties, and ultimately, if undertaken effectively, allow communities to accept and contribute positively to the proposed activity. Public consultation can lead to a negative decision on the project, when it is found that the activity will adversely affect local populations beyond acceptable means. Another factor unique to the Arctic EIA process is the possible "transboundary" impacts which can occur with activities. EIAs can be difficult as project developers and authorities need to make allowances for different legal systems, and matters are further complicated as the citizens within the area of likely impacts of the affected country need to be given the opportunity to also participate in the EIA. In the past, activities in the Arctic, which have led to "transboundary" impacts, have included development of oil and gas resources, large-scale hydroelectric projects, and road building. Clearly, bi- or multilateral agreements are required.

Many would ask whether the current Arctic legal regime is protecting the environment, and some would say that the answer depends on what nation is studied and what environmental issue is considered. The consolidation of the Arctic legal regime into a region-wide treaty, like the Antarctic Treaty, has been considered, and there are of course pros and cons. The main advantage in terms of EIA would be increasing states' obligations to protect the environment; however, time and expense of developing an international treaty which encompasses all the specific needs of the Arctic, such as the socioeconomic and cultural needs of the indigenous peoples, balanced against the unique environmental attributes of the Arctic, is likely to be a difficult task.

6.12 Conclusion

Five case studies were investigated to assess the effectiveness of the Environmental Impact Assessment system with respect to predicting and managing the effects of visitor activity in the Ross Sea region of Antarctica. Investigated EIAs included the decommissioning of the Greenpeace World Park Base, Cape Evans; Taylor Valley Visitor Zone, Taylor Valley; the decommissioning of the Vanda Station; occupation of the Loop Moraine campsite, both in the Wright Valley, McMurdo Dry Valleys; and the occupation and restoration of the ice-free storage area for the Cape Roberts Drilling Project, Cape Roberts. The assessed sites were situated on a variety of landforms, soil parent materials, and local climates. At each site, the history of the site was known, and the time since last disturbance was well constrained. Assessment of the present state of the sites and accuracy of the EIA predictions compared with the actual impacts were carried out in the austral summers of 2008/2009 and 2009/2010 using visual soil assessment methods.

Each of the five case studies demonstrates a high standard of EIA compliance, and appropriate remediation of sites post-event and follow-up monitoring where promised. Cleanup practices were of a high standard in each of the case studies investigated. The raking out of tracks, redistribution of disturbed stones, and careful backfilling of excavations all assisted in the rehabilitation of the formerly disturbed sites so that former occupation was almost undetectable at all camp sites investigated. In all cases, recovery was a function of the effort of the personnel doing the cleanup and active wind-, water-, and freeze-thaw-driven surface processes.

On the exterior, the EIA process may seem superficial with a lot of generic form filling. However, the great benefit of the EIA system is its ability to raise environmental awareness as visitors carry out their activities. The EIA system motivates visitors to take responsibility for their "footprint" on the environment. Thus, visitors are more likely to avoid creating impacts and are motivated to restore disturbed surfaces as much as possible prior to their departure. The result is an exceptionally high, world leading, standard of environmental management that reflects the relatively pristine environment in the Ross Sea region of Antarctica.

Uncertainties remain with regard to cumulative impacts. The VISTA (visitor site assessment scheme) is attempting to provide ongoing monitoring in the Taylor Valley Visitor Zone and building baseline information on recovery between seasons. EIA follow-up investigations such as this study are often a neglected element of the EIA process. Effort should be made to systematically carry out evaluations on a regular basis to ensure the EIA process continues to work effectively.

In contrast to the pristine Antarctic, the higher level of human occupation and activity in the Arctic means that EIA practices there are applied only to major development activities that are on a scale not yet contemplated in the Antarctic.

Acknowledgements Thank you to Antarctica New Zealand for logistics support over the summers of 2008/2009 and 2009/2010. We also thank Landcare Research for support, particularly the Murray Jessen Memorial Doctoral Scholarship through which Tanya O'Neill is funded. We would also like to thank Errol Balks for field assistance, Margaret Auger and Nathan Cross for science support at Scott Base, and Jana Newman and Ceisha Poirot for technical advice.

References

Abaza H, Bisset R, Sadler B (2004) Environmental impact assessment and strategic environmental assessment: towards an integrated approach. UNEP, Geneva

Antarctica (Environmental Protection) Act (1994) Public Act 1994 No. 119. Act is administered by the Ministry of Foreign Affairs and Trade, New Zealand

Antarctica New Zealand (2009) VISTA update and review: visitor site monitoring in the Ross Sea region. Antarctica New Zealand

Antarctica New Zealand (2011a) Statement of intent 2011–2014. The New Zealand Government

Antarctica New Zealand (2011b) Historic hut and Cape Roberts visitor statistics. Antarctica New Zealand

Campbell IB, Claridge GCC (1987) Antarctica, soils, weathering processes and environment. Elsevier, Amsterdam

Campbell IB, Keys JR (1993) Cape Roberts comprehensive environmental evaluation – a survey of the soils and vegetation. Unpublished report

Campbell IB, Balks MR, Claridge GGC (1993) A simple visual technique for estimating the impact of fieldwork on the terrestrial environment in ice-free areas of Antarctica. Polar Rec 29(171):321–328

Campbell IB, Claridge GGC, Balks MR (1994) The effects of human activities on moisture content of soils and underlying permafrost from the McMurdo Sound region, Antarctica. Antarctic Sci 6(3):307–314

Campbell IB, Claridge GGC, Balks MR (1998) Short- and long-term impacts of human disturbances on snow-free surfaces in Antarctica. Polar Rec 34(188):15–24

Claridge GGC, Campbell IB, Powell HKJ, Amin ZH, Balks MR (1995) Heavy metal contamination in some soils of the McMurdo Sound region, Antarctica. Antarctic Sci 7:9–14

Codling R (1995) The precursors of tourism in the Antarctic. In: Hall CM, Johnston M (eds) Polar tourism-tourism in the Arctic and Antarctic regions. Wiley, Chichester, pp 167–177

Committee for Environmental Protection (2005) Guidelines for environmental impact assessment in Antarctica, resolution 4. CEP handbook, 2007

Greenpeace (1986) Greenpeace Antarctic programme: environmental assessment for 1986–97 season. Greenpeace International, Amsterdam

Greenpeace (1991a) Initial environmental evaluation, removal of World Park Base, Cape Evans. Greenpeace International, Amsterdam

Greenpeace (1991b) Greenpeace 1990–1991 Antarctic expedition report. Greenpeace International, Amsterdam

Greenpeace (1992) Greenpeace 1991–1992 Antarctic expedition report. Greenpeace International, Amsterdam

Greenpeace (1993) The Greenpeace report of the Antarctic environmental impact monitoring programme at World Park Base 1991–1992. Greenpeace International, Amsterdam

Greenpeace (1994a) Greenpeace 1992–1993 Antarctic expedition report. Greenpeace International, Amsterdam

Greenpeace (1994b) World Park Base Antarctica 1987–1992: treading lightly: a minimal impact Antarctic station. Greenpeace International, Amsterdam

Greenpeace (1997) Greenpeace 1996–1997 Antarctic expedition report. Greenpeace International, Amsterdam

Hayward J, Macfarlane MJ, Keys JR, Campbell IB (1994) Decommissioning Vanda Station, Wright Valley, Antarctica: initial environmental evaluation, March 1994
Keys JR, Barrett P, Cooper A, Davey F, McFarlane M, Preston D, Pyne A, Webster B (1994) Final comprehensive environmental evaluation: Antarctic stratigraphic drilling East of Cape Roberts in Southwest Ross Sea, Antarctica, prepared and written for the International Steering Committee for Environmental Protection. Department of Conservation, New Zealand
Mear R, Swan R (1987) In the footsteps of Scott. Jonathan Cape, London
Nowlan L (2001) Arctic legal regime for environmental protection. IUCN, Gland
Preliminary Environmental Evaluation K123 (2004) Preliminary environmental evaluation for K123: environmental protection of soils of the Ross Sea Region, prepared for Antarctica New Zealand, June 2004
Reich RJ (1980) The development of Antarctic tourism. Polar Rec 20:203–214
Roura R (2004) Monitoring and remediation of hydrocarbon contamination at the former site of Greenpeace's World Park Base, Cape Evans, Ross Island, Antarctica. Polar Rec 40(212):51–67
Sheppard DS, Campbell IB, Claridge GGC (1993) Contamination in the soils at Vanda Station. Institute of Geological and Nuclear Sciences report to Antarctica New Zealand (unpublished)
Sutherland D (2002) Lake Vanda monitoring report no. 7. National Institute of Water and Atmospheric Research Ltd, prepared for Antarctica New Zealand
Tin T, Fleming L, Hughes KA, Ainsley DG, Convey P, Moreno CA, Pfeiffer S, Scott J, Snape I (2009) Review: impacts of local human activities on the Antarctic environment. Antarctic Sci 21:3–33
Webster J, Webster K, Nelson P, Waterhouse E (2003) The behaviour of residual contaminants at a former station site, Antarctica. Environ Pollut 123(2):163–179

Antarctic Treaty Consultative Meeting Documents Referenced

IP = Information Paper
WP = Working Paper

IAATO (2010) IP 113 ATCM XXXIII: IAATO overview of Antarctic tourism: 2009–10 season and preliminary estimates for 2010–11 and beyond
New Zealand, United Kingdom, United States (2006). WP 34 ATCM XXIX: Ross Sea Protected Area Inspections 2006

Websites

Environmental Impact Assessment statement at http://www.antarcticanewzealand.govt.nz. Accessed 14 Feb 2011
Logistics data at http://www.comnap.aq. Accessed 14 Feb 2011
Tourism Statistics at http://www.iaato.org. Accessed 14 Feb 2011

Chapter 7
German Tourism Activities in the Antarctic Area: A Governmental Perspective

Manuela Krakau and Heike Herata

Abstract The interrelationship between tourism development and polar environments became a hot topic in the last decades. Since climatic changes lead to altered ecosystems including the alteration of species, composition in areas with previous icy and hostile conditions, especially the use of polar landscapes, came into touristic focus. However, the Southern Ocean surrounding Antarctica has been an exceptional destination to experience nature and follow the path of adventurers since its discovery. To fulfill the Protocol on Environmental Protection to the Antarctic Treaty (Environmental Protocol) with the aim to protect the Antarctic area as well as possible, the procedure of granting a permit for tourism activities is well developed in German legislation. For decades, Germans have accounted for more than 10% of all tourists enjoying the special landscape during a cruise to Antarctic coasts. This puts them in the top 3 of the world together with American and British tourists. Since 2000, the German Federal Environment Agency recorded information on German tourist activities in the area of Antarctica from yearly post-visit reports. These data may support discussions on possible steps toward required regulations for polar tourism in the near future. Some aspects of land-based tourism on the Antarctic continent, and especially on the Antarctic Peninsula and islands nearby, are highlighted with regard to the potential consequences within the frame of environmental protection and climate change. Due to the lack of knowledge which impedes thoroughly environmental assessments for all touristic activities, the German Federal Environment Agency aims to encourage research and political institutions to support the development of a concept of sustainable tourism for the Antarctic Treaty area.

Keywords Tourism • Antarctic environment • Ecosystem • Antarctic Treaty • Germany

---

M. Krakau, Ph.D. • H. Herata (✉)
German Federal Environment Agency, Postfach 1406, 06813 Dessau-Roßlau, Germany
e-mail: Manuela.Krakau@uba.de; heike.herata@uba.de

D.K. Müller et al. (eds.), *New Issues in Polar Tourism: Communities, Environments, Politics*, DOI 10.1007/978-94-007-5884-1_7,

7.1 Tourism Development and the Antarctic Environment

Antarctica has been a destination for adventurers since its discovery in the eighteenth century. After the peak of commercial interests like whaling in the Southern Ocean, humans started to explore the white desert for fame and glory. Since the mid-twentieth century, tourists visiting the sixth continent have been fascinated by the special fauna and landscape. However, concern over melting glaciers and the appearance of nonnative species on the continent has been raised (IPCC 2007; Turner et al. 2009). Thus, the Antarctic continent seems to be threatened by man-made changes which are not yet assessed on the whole (Turner et al. 2009; Aronson et al. 2011).

In December 1966, the first journey to the Antarctic was organized by Lindblad for 58 passengers using the Argentine marine ship “Lapataiain” (Headland 1994). In 1971, a specially constructed cruise ship, the “Explorer,” was brought into service by Lindblad to let more people discover the great nature of the Antarctic Peninsula. This was the start of organized tourism inviting several hundreds of people per season to Antarctica. In the last decade, up to 40 international cruise ships carrying more than 40,000 tourists visited the white continent yearly from November to March (IAATO’s yearly tourism statistics, also in, e.g., Cessford 1997). The availability of Russian ice breakers in the 1990s, previously used in scientific surveys, probably enhanced that development (IAATO Tourism statistics 2000–2001). In these times, half of the Antarctic cruise vessels bore Russian names (9 out of 18 ships). Most Antarctic tour operators are organized in the “International Association of Antarctic Tour Operators” (IAATO) founded in 1991, and the majority of people come by cruise ships which have strong interest to keep Antarctica as a pristine tourist destination as stated in the IAATO objectives (IAATO 2011).

Yachts are less well recorded in appearance and movement than cruise ships. Private yachts may cause impairment of the environment because some of them visit Antarctic waters without appropriate knowledge of the rules of best practice and without any permit. Small boats navigated by sailors inadequately prepared for polar challenges may display a risk to themselves but also to the rescue teams (e.g., resulting in sinking of BERSERK, described by New Zealand et al. (2011)). Thus, the initiative of IAATO to elucidate risks of Antarctic cruising for private sailors (IAATO information for yachts 2011) was highly welcomed.

Until the crash of a New Zealand plane in 1979, touristic flights over the Antarctic Treaty area made up a small part to the tourism industry (IAATO Tourism statistics). For several years, no attempts were started to revive the overflight business again. Compared to ship cruises, there is still only low interest in such flights. But combined touristic offers of flight and cruise packages (“air-cruise program”) started to increase in the last years (IAATO Tourism statistics). Land-based tourism, mainly private expeditions to interior of the white continent, has been enhanced by support of Adventure Network International since 1984 (Splettstoesser and Folks 1994).

7.2 Antarctic Tourism and International Legal Regulations

Although tourism in the Antarctic Treaty area is under discussion at "Antarctic Treaty Consultative Meetings" (ATCM) for years, the sinking of the MV "Explorer" in 2007 pushed forward recent debates on binding regulations. After several voluntary regulations, the discussion on legally binding rules for construction, equipment, and operational improvements came into focus within the relevant bodies like ATCM and the International Maritime Organization (IMO). Although ATCM Measure 4 (2004) early pointed out the importance of contingency planning and insurance issues, it is not implemented in all consultative parties' national law and is thus not yet in effect. The regulation of tourism impacts on landing sites was considered in Measure 15 (2009) and is based on existing IAATO requirements for cruise ships. Locally restricted visitor site guidelines have been regularly implemented as resolutions since 2005 to protect Antarctic environments frequently visited by tourists.

Regulations of ship-based tourism mainly concern security aspects but also consider the protection of sea and coast of that "special area" (cf. IMO's International Convention for the Prevention of Pollution from Ships, MARPOL 73/78, Annexes I (oil), II (noxious liquid substances), and V (ship-generated garbage)). An important step toward the protection of polar waters may be IMO's resolution within MARPOL on the ban of heavy fuel onboard ships cruising in the Antarctic Treaty area (Chapter 9, Annex I of MARPOL). This regulation was supported by ATCM Resolution 7 (2010) and became effective in August 2011. Recommendatory "Guidelines for Ships operating in Arctic and Antarctic ice-covered waters" were considered by ATCM Decision 4 (2004) and by IMO through Resolution A.1024(26) in 2010. Ship safety and prevention of environmental pollution are under review and will be gathered in the mandatory Code of Safety for Ships operating in Polar Waters (Polar Code) developed by IMO and supported by the Antarctic Treaty Parties (ATCM Resolutions 3 (1998) and 8 (2009)). The recent introduction of IAATO's self-regulation of vessel tracking with hourly reporting for all IAATO SOLAS passenger vessels (since season 2010/2011) will enhance search and rescue operations (IAATO 2009). The ban of open lifeboats by October 2010 and their substitution represents an important step toward human safeguard. The tightening of safety-related requirements for lifeboats for polar waters is also under discussion within IMO and supported by ATCM (cf. Resolution 8 (2009)).

Although mentioned in ATCM Resolution 1 (2003), private yacht cruises are the kind of tourism in Antarctic waters that is least paid regard to. After the sinking of the Norwegian "Berserk" in season 2010/2011, henceforth the focus is on improving information of private sailors and the regulation of technical requirements and training (United Kingdom 2011; Germany et al. 2011; IAATO information for yachts 2011). The fact that this vessel had no authorization to visit the Antarctic Treaty area, like the French yacht the year before, entails the question on control and penalty feasibility (cp. France 2011; Norway 2011). Nevertheless, the implementation of ATCM regulations in national law of such rules may vary and span differently long periods. The differing interpretation of the Environmental Protocol leads to varying practice of permitting processes, e.g., for yacht cruises, even within consultative parties of the Antarctic Treaty. The regulation of technical requirements and training demands may become a hot topic the next years in the face of climate change and melting ice, thus generating new attractive sailing sites. Furthermore, it could be useful to apply, to a large extent, the above mentioned polar regulation system to those vessels which do not sail under the flag of any of the Antarctic Treaty Parties. This approach could possibly prevent an increase of unregulated visits.

7.3 Authorization Procedure for Tourism Activities in Germany

Pursuant to the Act Implementing the Protocol on Environmental Protection to the Antarctic Treaty (AIEP 1998), any activity whether related to scientific research, tourism, or journalism that is organized in Germany or begins on German territory is subject to authorization. As the national competent authority, the Federal Environment Agency (Umweltbundesamt, UBA) assesses the impact of the planned activity on the Antarctic environment and the ecosystems that are associated with or depend on it. UBA was already founded in 1974 with the goal to support the government on scientific level, to enforce the environmental law, and to inform the public on topics regarding environmental protection. Work topics include in general a high variety of environmental aspects. Besides planning of landscape usage and sustainable strategies, health care and ecosystem protection, sustainable techniques, chemical security, and the regulation of emission trading are in the responsibility of UBA.

According to the AIEP, individuals as well as commercial tour operators have to submit to UBA an application for their planned activity. To get all information necessary for the assessment of this activity, specially designed questionnaires are easily available on UBA's website (Fig. 7.1). Before taking a decision to issue a permit, UBA asks the Federal Maritime and Hydrographic Agency (Bundesamt für Seeschifffahrt und Hydrographie, BSH) on possible marine hazards arising from the planned activity. BSH evaluates constructional and technical data of the ship and checks the validity of ship certificates. In addition, this agency looks at the planned itinerary to estimate potential impacts to the marine environment arising from the navigation through ice-covered waters of the Antarctic. Thus, e.g., the ice class of a ship may lead to restrictions of regional destinations for landings of tourists in Antarctica. Planned landings and detailed descriptions of activities onshore are crucial for UBA's environmental impact assessment. On the basis of an appropriate study demanded from the operators of cruise ships, UBA carries out an Initial Environmental Evaluation (IEE) for each ship and each

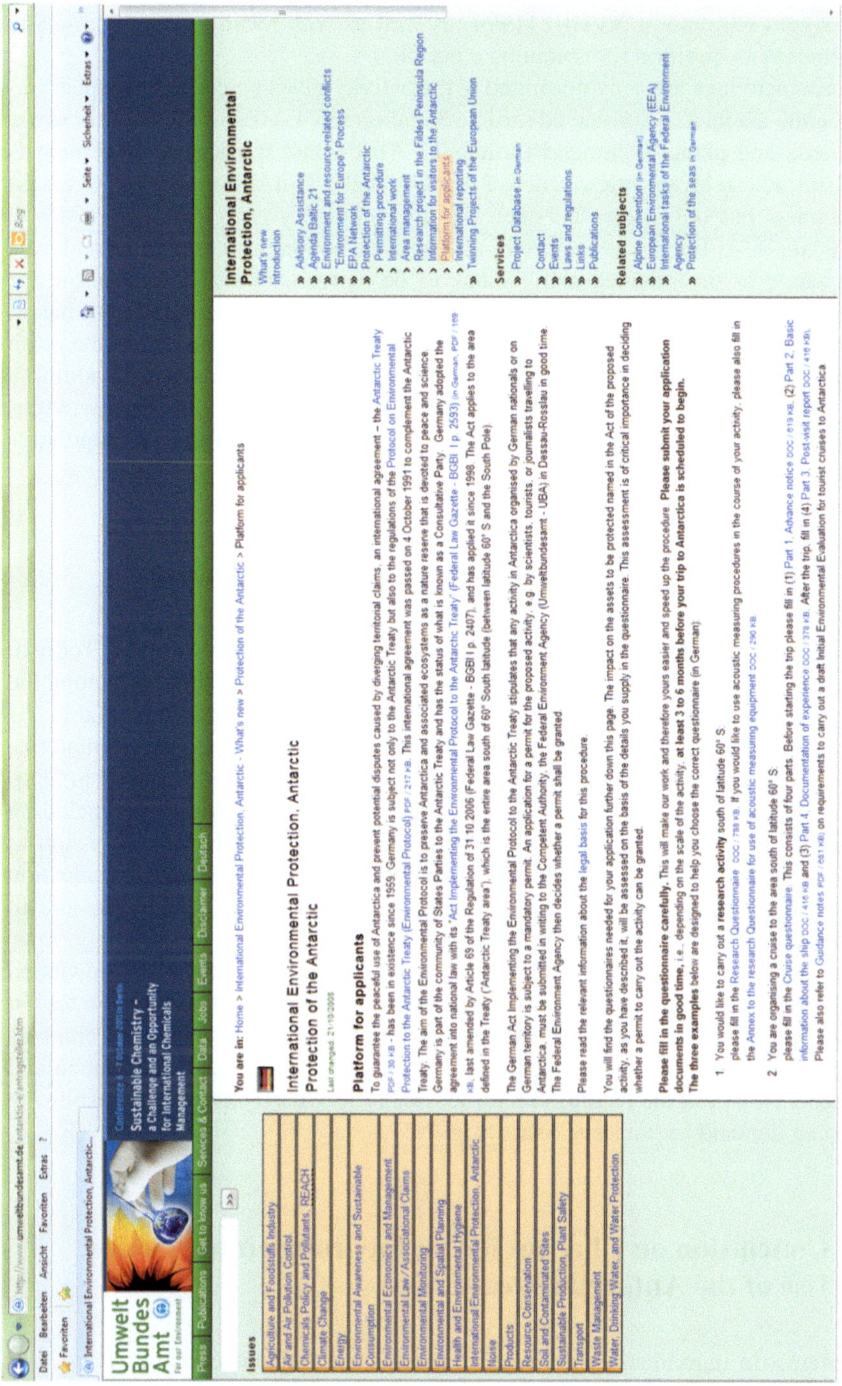

Fig. 7.1 German Internet platform for people intending to travel to the Antarctic Treaty area including several links to the legal background

season. The strict compliance with the German Guidelines for Visitors according to the Kyoto Resolution XVIII-I (1994) as well as with locally valid Visitor Site Guidelines is a condition for obtaining a permit.

Tourist activities are only permitted if the activity gives no cause to suspect impacts on the assets to be protected (article 3, paragraph 4 AIEP), if the conservation of animals and plants is granted (article 17 AIEP), and if waste management is compliant with relevant rules (article 21 to 24 AIEP). If it is not possible to assess real impacts but only potential ones, the precautionary principle is considered. Permits are valid for one season only and have to be renewed yearly. In addition, each change in the planned activity has to be announced to UBA to allow for reevaluating possible changes in conditions. After the visit, each applicant has to submit a report to show that the activity was carried out in accordance with the given rules and requirements and to record potential extraordinary incidents. In addition, these reports provide the basis to gather information of the different activities as background of future assessments and potential consideration of cumulative impacts.

7.4 Review of German Tourism over the Last Decade

Germans account for more than 10% of all tourists to Antarctica (IAATO Tourism statistics). Together with the American and the British travelers, they are among the top 3 of nations visiting the Antarctic (IAATO Tourism statistics). As Table 7.1 illustrates, a high number of German tourists were recorded visiting Antarctica from the period of 2006 to 2009. A noted decline of visitation occurred in 2010 (Table 7.1). However, numbers of German tourists traveling to Antarctica including journeys on other nation's ships tell a slightly different story which is more similar to the international trend recorded by IAATO (Fig. 7.2). IAATO's data collection complements the post-visit reports submitted to UBA and makes it possible to profile German activities in the Antarctic.

Tour operators' interest to show intact nature and endemic fauna has grown and concentrates on the Antarctic Peninsula. Between 2004 and 2008, German tourist mainly visited the northern tip and the western part of the Antarctic Peninsula (Fig. 7.3). Most of German ships headed for islands with extraordinary fauna and landscape. However, the landings on these islands decreased only a little despite the increasing demand for touristic visits (Fig. 7.3).

7.5 Conclusion on Chances for an Environmentally Sound Use of the Antarctic Treaty Area

The interest in individual exploration of the white continent is rising. Thus, it is not surprising that new branches of Antarctic tourism develop. Along with the changing number of tourists, the interest and expectations of tourists altered over the

Table 7.1 German statistics on authorized cruise ships per season

Season	2001/2002	2002/2003	2003/2004	2004/2005	2005/2006	2006/2007	2007/2008	2008/2009	2009/2010	2010/2011
Number ships	3	2	3	3	2	6	5	6	6	3
Nr ships >200 pax	1	0	1	1	0	4	2	2	3	0
Trips in total	15	11	16	12	11	15	19	20	20	12
Pax in total	2,006	138	2,407	1,906	1,445	3,481	3,135	4,163	3,214	1,168

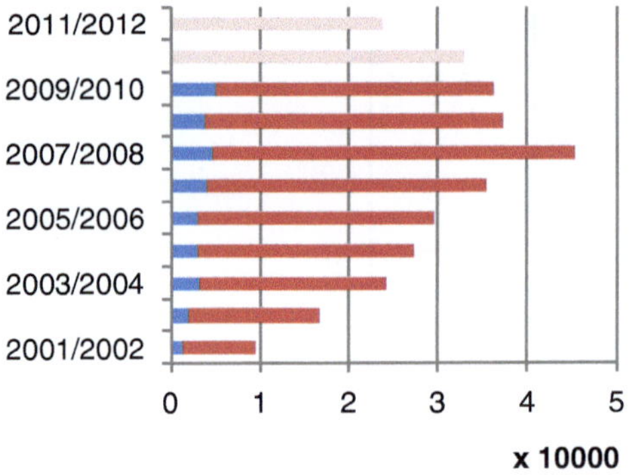

Fig. 7.2 Number of German tourists in relation to tourists in total on IAATO ships (Data source: IAATO Tourism statistics). The bars for 2010 and 2011 were estimated by IAATO

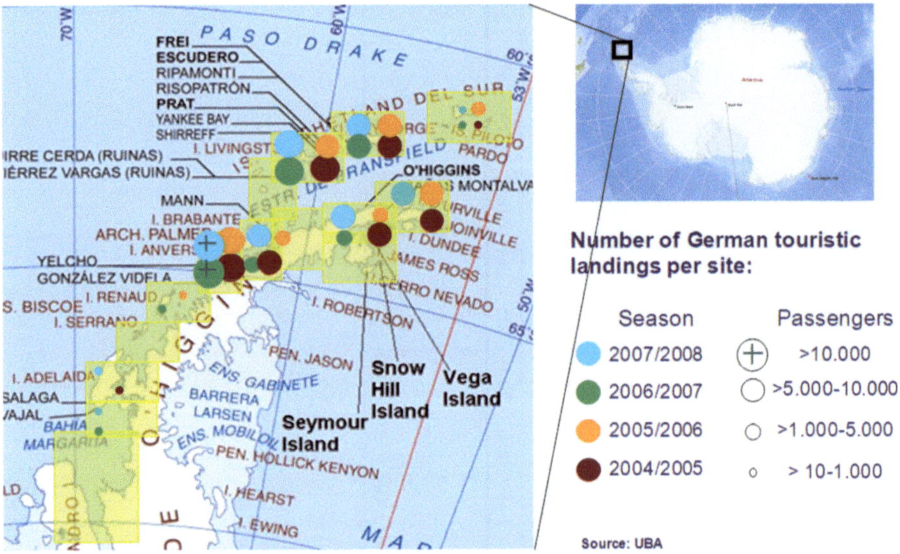

Fig. 7.3 Preferred landing sites of German cruise tours over a 4-year period (Data were extracted from post-visit reports to the German Federal Environment Agency)

years. Thus, there is a change from observing trips aboard cruise ships to intensified experience and adventure tourism including kayaking and camping which needs more consideration (IAATO 2003; United Kingdom 2004; Russian Federation 2008). Offers of extreme sport events, adventure trips, and stays overnight are growing as well as yacht tourism and air-cruise voyages. Here, the German Federal Environment Agency aims to inspire research off the cruise ship tracks. The influence of private expeditions on Antarctic pristine regions and the interference with scientific surveys are to be estimated to add that information to the needed whole ecosystem assessment.

Various possible developments of land-based tourism on the Antarctic continent, and especially on the Antarctic Peninsula, are under discussion at ATCMs with regard to the potential consequences within the frame of environmental protection and progressing climate change. But also cruise tourism will face the changes in polar areas, e.g., by increased access due to sea ice retreat or adopted itineraries due to attractive species' local shifts (Lamers and Amelung 2010). The development and consequent implementation of site guidelines are strengthened for areas of high attraction for tourists, and a generic guideline applicable to any landing site was created (ATCM Resolution 3 (2011)). The idea is to prevent overuse and destruction of the corresponding site. One example of such a successful management is Hannah Point where a complete closing of the landing site helped the penguin colony to recover from strong losses due to tourist visits. Today, only restricted access is possible to that area (ATCM site guidelines).

The regulation of cruise ships was and will be influenced by the decision of international bodies, like IMO which currently develops the Polar Code. In future seasons, the composition of the Antarctic tourism fleet will be affected by IMO's resolution on the ban on the use and carriage of heavy fuel oil (MEPC 189 (60)) which came into effect in August 2011. However, a big change in the Antarctic tourism sector is not expected since expedition cruises are supposed to represent the major part in future, too. According to IAATO data, cruise ship tourism seems to have reached its maximum in 2007/2008 and will probably stabilize at about 20,000 guests per season visiting the Antarctic continent or islands nearby.

Considering the idea of protection of the Antarctic, it is discussed how cruise ships as well as individual activities can be framed to follow a concept of sustainable tourism (cp. Johnson 2002, Lamers 2009). These questions may be answered by a visionary concept that should be developed by all parties involved. To date, there is an obvious scarcity of data on the common motivation of tourists to visit Antarctica and on the suspected trend that nature becomes less attractive than sport events. In addition, it would be useful to have better estimates regarding the ability of different landing sites to cope with strains from tourist visits. In this connection, the changes of landscape and responses of biota to climatic variations have to be kept in mind. Therefore, the development of a special kind of doomsday tourism should be prevented.

In general, the German Federal Environment Agency worries about a change from well-regulated cruise tourism to less-manageable individual activities. So far, it is difficult to track the ways of individual adventurers and to encourage their environmentally sound behavior. Moreover, consequences of extreme sport, like marathon events, climbing challenges, or rock jumpers, are difficult to assess since those activities strongly depend on the area chosen and on its accessibility as well as the number of participants and supporting crew. Here, UBA prefers the restriction or avoidance of land-based adventure tourism to protect the Antarctic wilderness values and preserve untouched areas for research only. To counteract the development of mass tourism in the Antarctic continental area, limitation of overnight stays should be enforced by a moratorium against the establishment of hotels in the Antarctic Treaty area. In addition, it will be necessary to keep an eye on the potential increase of flights, which is already observed by IAATO since DAP (Chile) and TAC (South Africa/Russia) are operating.

Long-term approaches have to be followed since there are too less-efficient regulations impeding tourism impacts. While research is privileged by the Environmental Protocol, neither tourism in its current dimension nor its future development was explicitly considered. An adapted additional annex for tourism might be an option to integrate the topic into the legal frame of the Environmental Protocol. Since such regulations require time for political discussion and national preparation for implementation, local regulations like zoning and Antarctic site guidelines seem to be useful tools for the next years. To react to short-term variation in climatic conditions and their consequences for the landscape, anticipatory regulation and site management are very important issues. With regard to conserve the pristine continent with its unexplored regions, the wilderness value has to be the first concern. In intermediate term, the German Federal Environment Agency deems best the implementation of measures to improve safety and environmental awareness and a moratorium on restricted land-based tourism.

References

AIEP (1998) The protocol on environmental protection to the Antarctic treaty

Aronson RB, Thätje S, McClintock JB, Hughes KA (2011) Anthropogenic impacts on marine ecosystems in Antarctica. Ann N Y Acad Sci 1223(1):82–107

ATCM Resolution 3 (1998) International code of safety for ships in polar waters. ATCM XXII – CEP I, Tromsø

ATCM Resolution 1 (2003) Advice to vessel and yacht operators. ATCM XXVI – CEP VI, Madrid

ATCM Decision 4 (2004) Guidelines for ships operating in Arctic and Antarctic ice covered waters. ATCM XXVII – CEP VII, Capetown

ATCM Measure 4 (2004) Insurance and contingency planning for tourism and nongovernmental activities in the Antarctic Treaty area. ATCM XXVII-CEPVII, Cape Town

ATCM Measure 15 (2009) Landing of persons from passenger vessels. ATCM XXXII – CEP XII, Baltimore

ATCM Resolution 8 (2009) Mandatory shipping code for vessels operating in Antarctic waters. ATCM XXXII – CEP XII, Baltimore

ATCM Resolution 7 (2010) Enhancement of port state control for passenger vessels bound for the Antarctic Treaty area. ATCM XXXIII – CEP XIII, Punta del Este

ATCM Resolution 3 (2011) General guidelines for visitors to the Antarctic. ATCM XXXIV – CEP XIV, Buenos Aires

Cessford G (1997) Antarctic tourism – a frontier for wilderness management. Int J Wilderness 3(3):7–11

France (2011) Follow-up to the unauthorized presence of French yachts within the treaty area and damage caused to the hut known as Wordie House. ATCM XXXIV, Buenos Aires, WP 011

Germany, Australia, Norway, United Kingdom, United States (2011) Yacht guidelines to complement safety standards of ship traffic around Antarctica. ATCM XXXIV, Buenos Aires, WP 037

Headland RK (1994) Historical development of Antarctic tourism. Ann Tour Res 21(2):269–280

IAATO (2003) Adventure tourism in Antarctica. ACTM XXVI, Madrid, IP 096

IAATO (2009) IAATO makes recommendations on Antarctic passenger vessel marine safety at 20th annual meeting in Providence, RI, USA. Press release from June 26, 2009

IMO (2010) Guidelines for ships operating in polar waters. Resolution A.1024(26). IMO, London

IPCC Fourth Assessment Report: Climate Change 2007: IPCC (2007) Climate change 2007: synthesis report. Contribution of working groups I, II and III to the fourth assessment report of the intergovernmental panel on climate change, Core Writing Team, Pachauri RK, Reisinger A (eds), 104 pp. IPCC, Geneva. http://www.ipcc.ch/publications_and_data/publications_and_data_reports.shtml

Johnson D (2002) Environmentally sustainable cruise tourism: a reality check. Mar Policy 26:261–270

Lamers M (2009) Sustainable tourism in Antarctica? Perspectives, policies and prospects. In: Lamers M (ed) The future of tourism in Antarctica – challenges for sustainability. Dissertation, University of Maastricht, pp 155–175

Lamers M, Amelung B (2010) Climate change and its implications for cruise tourism in the polar regions. In: Lück M, Maher PT, Stewart E (eds) Cruise tourism in polar regions – promoting environmental and social sustainability? Earthscan, London, pp 147–163

New Zealand, Norway, United States (2011) The Berserk incident, Ross Sea, February 2011. ATCM XXXIV, IP 018

Norway (2011) The legal aspects of the Berserk expedition. ATCM XXXIV, Buenos Aires, IP 075

Russian Federation (2008) Proposals for regulating the adventure tourism and non-governmental activity in the Antarctic. ATCM XXXI, Kyiv, WP 050

Splettstoesser J, Folks MC (1994) Environmental guidelines for tourism in Antarctica. Ann Tour Res 21(2):231–244

Turner J, Bindschadler R, Convey P et al (2009) Antarctic climate change and the environment. Scientific Committee on Antarctic Research, Cambridge

United Kingdom (2004) Managing adventure tourism: the need for enhanced co-operation amongst parties. ATCM XXVII, Cape Town, WP 002

United Kingdom (2011) Training course for yachts intending to visit Antarctica. ATCM XXXIV, Buenos Aires, IP 015

Website References (Accessed October 2011)

ATCM site guidelines: http://www.ats.aq/e/ats_other_siteguidelines.htm

IAATO tourism statistics: http://iaato.org/de/tourism-statistics

IAATO 20th annual meeting decision on vessel tracking: www.iaato.org/press/IAATO_Actions_to_Enhance_Marine_Safety.pdf

IAATO objectives: http://iaato.org/de/objectives

IAATO information for yachts 2011: http://iaato.org/de/yachts

Kyoto Resolution XVIII-I (1994) UNFCCC, New York

UBA visitor guidelines: http://www.umweltbundesamt.de/uba-info-medien-e/search-public-e.php, search for: visitor guidelines

UBA platform for applicants: http://www.umweltbundesamt.de/antarktis-e/antragsteller.htm

Chapter 8
The Wild North: Network Cooperation for Sustainable Tourism in a Fragile Marine Environment in the Arctic Region

Sandra Magdalena Granquist and Per Åke Nilsson

Abstract Rapidly increasing tourist flows to the Arctic North have put focus on the sustainable development of tourism in natural environments. The Wild North project shows how different stakeholders in a network across the Arctic North deal with the need for a development of sustainable use of wildlife in a tourism context by using interdisciplinary research, active cooperation, and exchange of knowledge. The wildlife focused in this international project consists of arctic foxes, whales, seals, and birds. The project demand came from the tourism industry and was taken up by public organizations and academies within nature protection and tourism management. The project has also the maintainability of a destination in focus together with its sustainability.

Keywords Wildlife tourism • Sustainability • Maintainability • Code of conduct • Triple helix

---

S.M. Granquist
Institute of Freshwater Fisheries, Brekkugata 2, 530 Hvammstangi, Iceland

The Icelandic Seal Centre, Brekkugata 2, 530 Hvammstangi, Iceland
e-mail: sandra@veidimal.is

P.Å. Nilsson, Ph.L. (✉)
Hólar University College, 551 Hólar, Sauðárkrókur, Iceland

The Icelandic Seal Centre, Brekkugata 2, 530 Hvammstangi, Iceland

Hammerdal Förlag & Reportage, Häradsvägen 11, 83070 Hammerdal, Sweden
e-mail: pan@hammerdalforlag.se

D.K. Müller et al. (eds.), *New Issues in Polar Tourism: Communities, Environments, Politics*, DOI 10.1007/978-94-007-5884-1_8,

8.1 Introduction

Rapidly increasing tourist flows to the Arctic North have put focus on sustainable development of tourism in natural environments. Studies of Arctic tourism have had focus upon both benefits and problems for the destinations. Johnston (1995) understands, for example, the appeal of Arctic and sub-Arctic environments for tourists but recognizes that because of the impact on local communities and environment, tourism must be controlled. Hull (2001) points at the complexity behind a decision by the Canadian government to open up Labrador for worldwide adventure travel and the policy behind it with regard to a balance between use and protection.

Lately, climate change has put even more focus on how to match the impact of increased number of tourists. Increased tourism may raise the number of more general tourists with different preferences, implying different age groups, less individually skilled tourists with higher tolerance to crowding, and tourists with special interest in non-wildlife components (Catlin and Jones 2010). Vannini et al. (2006) take a wider perspective and points to the impact of general better access to peripheral areas by referring to the opening of the Northwest Passage, which will make it possible to reclaim a lost archipelago image in Canada but also give a complex socially and materially ecological situation.

Wildlife tourism has often been considered to be tourism with a minimum of impact activity on nature. Most of the discussions among tourism researchers have hence focused on how to plan for expanded use of nature and wildlife in a sustainable way (Hall 2008), how to meet ordinary people's steadily expanding need for access to nature and wildlife in accordance with preservation needs (Nilsson 2002), how to combine this expanded access to nature and wildlife as an inherent component of a vital democracy (Arler 2002), how to model plans for maintaining accessibility to nature in the shadow of increasing interests for wildlife experiences (Bosetti et al. 2009), how to establish environmental protection beneficial for generations of people ahead but also in the long run for industry and nature (Gössling and Hall 2006; Nilsson 2008), and how to construct guides for behavior of people in a nature environment where use and protection can meet the exceeding needs from different interests (Mason 1997; Grant 2002; Hughey et al. 2004; Garrod and Fennell 2004; Fennell 2007).

Growing interest for watching animals in their natural environment intersects, however, with a growing awareness of the potential humans display to disturb wild animals since it is well documented that human disturbance may affect the well-being and fitness of wild animals. Whoeler et al. (1994) find, for example, that overuse together with a growing interest for animal watching in some cases may reduce reproduction success as well as lead to changes in distribution of animals. Johnson and Lavigne (1999) identify tourism as among the most significant causes of the decline affecting monk seals in the Mediterranean Sea. Cassini (2001) finds in a study of fur seals in South America that distance between watchers and seals has importance: a closer distance than 10 m between tourists and seals caused a strong negative response among the animals. Orams (2002) describes how feeding wild animals as a tourist attraction causes damage among the animals since they adopt a change of behavior from hunting to being fed. Kent and Crabtree (2008) tested the effectiveness of establishing a sanctuary zone for sea lions to reduce human impact and found out that the sanctuary zone was ineffective, since the sea lions hauled out to a higher extent in a nearby recreational zone, where environmental factors were more suitable for hauling out. Since environmental factors often vary between years, a fixed sanctuary zone may therefore not be a good solution when aiming for reducing human impacts. Kovacs and Innes (1990) found that tourism affected the interactions between harp seals and their pups negatively, for example, leading to less female attention to the pups, as well as the females spending less time nursing their pups. Briassoulis (2002) claims that wise management of natural or constructed resources for tourist use may result in use by local residents as well, which in turn results in overused and degraded resources, severely threatening sustainable development. But even if he recognizes negative impact on marine environment, Lück (2007) has a hope that example of good practice and watchdogs may change business practice in a sustainable way.

This chapter shows how a network of tourism entrepreneurs, scientists, and public organizations in four North Atlantic countries in 2008 got an impetus to develop a sustainable use of nature in a tourism context by exchange of knowledge and expertise in the wildlife tourism sector. This has been done within a project called The Wild North where different actors according to a triple helix model (academy-industry-authorities) have participated. The fundamental questions raised in the project are how to unite the divergent interest of a booming tourism industry with a sustainable management of wildlife as a natural resource.

8.2 Theory

The concept sustainable development has been used quite extensively within tourism planning and policy making over the last few decades. According to Hall and Lew (1998), the concept of sustainable development first reached public attention when the World Conservation Strategy (WCS) was published at the Rio conference in 1980. The strategy should be formed by government agencies, individual experts, and nongovernmental organizations worldwide with the aim to make the environment "sustainable" for the twenty-first century. The word "sustainable" became spread by the so-called Brundtland Report Our Common Future in 1987, where sustainable development was defined as a development that meets the needs of the present use of environment without compromising with the ability of future generations to meet their needs. The expression "future generations" alluded to children born in the twenty-first century, and needs meant the wish that the state of environment today should also be at present for those children (World Commission on Environment and Development 1987:24). Although the concept has since then been widely recognized, it has caused challenges, whether it should be a conservation of the actual state of the art or contain some form of development within the frames of a sustainable environment.

To meet the need for such a sustainable development, focus has been more and more on cooperation and networking, seeking a balance between present benefits and increased future opportunities for the host community. This has led to concepts such as stakeholder theory (Freeman 1984) and collaboration theory (Gray 1989; Jamal and Getz 1995). According to Timur and Getz (2002), collaboration between stakeholders requires a high degree of interdependence, mutual benefits to be derived from the collaboration and that decisions arrived to will be implemented, encompassing public organizations, tourism industry, resident organizations and special interest groups, a convenor to initiate and facilitate the collaboration, and effective community process for strategic tourism planning with a vision, goals, and a self-regulation mechanism.

Another model used in this particular project is the triple helix model which highlights the interconnection between academy, government, and industry. Leydesdorff (2005) calls this a network of reflexive relations that adds surplus value by reorganizing and harmonizing the political and economic structures in order to achieve a better approximation of the variety of uncertain goals.

The big undertaking is to demonstrate the mutual benefits for all partners involved. There must be both sustainability and maintainability as an outcome of the project and that can be achieved if both tourism industry and research can agree upon how to manage the situation. Hardy and Beeton (2001) argue that maintainable tourism exists when the status quo is managed to deal with concerns related to short-term trends and impacts rather than to address the underlying fundamentals of sustainability. The views of major stakeholder groups like host population, tourists, and tourism industry must meet the needs of natural environment in a way to achieve balanced tourism (Ritchie and Crouch 2003). For tourism research with strong interests from end users but also with strong views on research paradigms from researchers where a positivistic view claims that natural science is "objective," it is essential to reflect how science is practiced and what makes it succeed or fail (Rodger et al. 2009).

8.3 Method

The Wild North project (TWN) started by setting up an ongoing monitoring of tourism impact to investigate in what way wildlife populations were affected by tourism activities. The results were presented to the tourism entrepreneurs so they could organize their operational procedures in a way that avoided a direct negative impact on local wildlife. Monitoring was regarded as important, since once negative impact on wildlife from tourism is noticed, it is often too late as the animal population already has been damaged beyond repair or relocated to another area (Ballantyne et al. 2009). On the other hand, monitoring is complicated and demands cooperation between the wildlife tourism industry and biological research groups (Orams 2002; Lück and Higham 2008; Strong and Morris 2010). The outcome of the project was aimed to develop local codes of conducts on how to behave around the wild animal species. Such codes have been a strategy for tourism management, even though their quality has not always been satisfying (Garrod and Fennell 2004; Quiros 2007; Curtin et al. 2009). Where such cooperation exists, it has been to a great benefit for environmental protection (Rodger et al. 2009). Research has also shown that a better understanding of the interaction between wildlife and tourism is necessary when looking toward a sustainable wildlife tourism development (Hall et al. 2003; Hughes 2002; Lück 2007).

8.4 The Wild North Project (TWN)

The main objective of the project was to contribute to sustainable development of wildlife tourism in a way that is beneficial for the communities/companies but still leads to a minimum effect on the wildlife. Preparations and partner search took 2 years, but a formal 4-year project funded by the North Atlantic Cooperation (NORA) and the North Atlantic Tourism Association (NATA) was started in 2009 with approximately 25 partners/stakeholders working together in seven clusters. A few partners also joined the project along the way after it had formally started. One cluster (Greenland) has abandoned the project.

8.4.1 Stakeholders and Clusters

The triple helix model became a natural way of organizing the process. The demand from the tourism industry together with actual environmental problems was discussed with different research centers.

The main method was to study the interaction between tourists and animals. The number of tourists at each cluster varied and so did the type of animal. Since the destinations were open for public, it was impossible to count the number of visitors. The number of animals was easier to estimate but of course not comprehensive. The research process was based on observations of interaction between tour entrepreneurs/tourists and direct effects on animals. Tourists were asked about their views on tourism in general and on the specifically watched animal.

In Iceland, there were three clusters. At Hornstrandir Reserve in the West Fiords, arctic fox study tours were followed and the human/animal interaction was studied. The tourists went to the island by boat and stayed in tent for one or more days. At Vatnsnes Peninsula, harbor seal watching was observed, and the interaction between tourists and seals was recorded. All tourists came by car and mainly consisted of small groups. At Húsavik, various species of whales were studied. Two companies sailed in the bay. Most tourists in Iceland were foreigners, mainly from Central Europe.

At Vestmanna in the Faroe Islands, birds were studied. The tourists were offered a boat tour by a private entrepreneur, and most of them were Scandinavians but also from Central Europe and the UK.

In Nuuk Fjord in Greenland, whale watching was managed by one entrepreneur and Nuuk tourist office. The number of tourists was rather low, and they all came by charter. At the end of the project period, the Greenland partner did not find sufficient benefit from the project and dropped out.

In Kvænangen in Norway, an island-based local entrepreneur organized both accommodation and bird watching tours, both by boat and by foot. A lot of the visitors came from Russia, and many came both for watching and hunting/fishing.

8.5 TWN Work Packages

When designing the project, the management group tried to address issues that partners found lacking in the partners' countries but still very important for a successful sustainable development of their wildlife tourism businesses and destinations. With this in mind TWN partners put together three work packages: research, education and knowledge transfer, and product development (see Fig. 8.1), aiming at a code of conduct and a business charter.

The TWN work packages were designed to fulfill the following objectives to:

- Estimate a baseline for wildlife populations at wildlife tourism destinations
- Evaluate the effect of tourism on wildlife populations
- Explore how tourists behave at wildlife destinations
- Explore how tourists perceive sustainable development at wildlife destinations
- Explore the background of tourists at TWN destinations
- Educate tourism operators and other working within wildlife tourism
- Eransfer knowledge between researchers, public organizations, and the tourism industry
- Develop sustainable wildlife tourism products

8.5.1 Work Package 1: Research

This work package consists of biology research and tourism research. Wildlife mapping and behavioral research of the animals was conducted at most destinations throughout the project period. Research was carried out by biologists, specialized in the relevant animal species at each location. Different methods relevant for the species in question were used to observe tourism impact (Granquist and Nilsson 2011). This was done to monitor the impact tourists have on the behavior and welfare of wildlife at TWN destinations. The results of the research have been and will be used to develop wildlife tourism at the different destinations, by creating a TWN Code of Conduct for tourism operators and tourists.

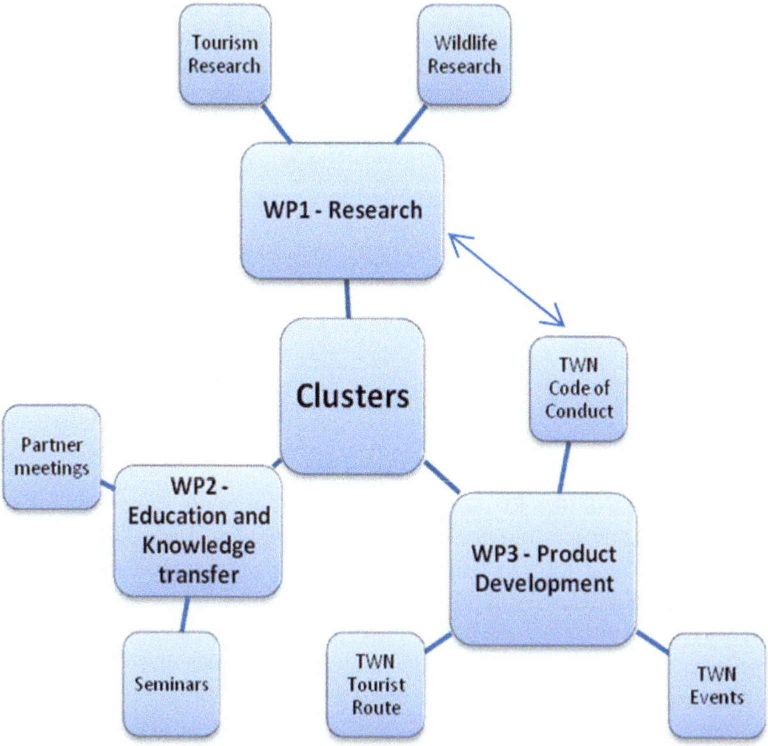

Fig. 8.1 The structure of TWN project where the cooperation clusters participate in most or all work packages through active cooperation

8.5.2 Work Package 2: Education and Knowledge Transfer

At annual seminars, directly aimed at tourism companies, guides and others, working closely within the wildlife tourism sector, various business practices towards a more sustainable industry and experiences were discussed and project activities and preliminary results were shared. These seminars aimed to be of practical assistance to those who attended them, making it easy to integrate the seminar recommendations and techniques into day to day practices.

8.5.3 Work Package 3: Product Development

Several products have been and will be developed within the project like The Wild North Code of Conduct (TWNCC), TWN Events, and the TWN Route. The project's end product will be the TWN accreditation scheme supported by TWNCC for guests at wildlife sites and a business charter for tourism companies. The TWNCC is based on research results from work package 1, and its objective is to minimize the negative effect of tourism on wildlife populations by educating both tourists and operators about proper behavior at wildlife tourism destinations. TWN will be established as an organization of accredited partners with ongoing monitoring of tourism impacts at wildlife destinations. During TWN Events visitors are invited to participate in some parts of the research process. The objective of these events is to introduce the TWN locations through unconventional methods creating positive publicity and broader target groups. An example of such an event is The Great Seal Count at Vatnsnes Peninsula (Iceland), where volunteers walk altogether100 km of coastline in 2–3 h counting seals along the way.

8.6 Discussions

The project is based on the ideology of sustainable tourism development and the partners' mutual understanding that wildlife is a valuable natural resource that must be maintained and that tourism entrepreneurs have a responsibility regarding the protection and management of wildlife tourism destinations.

The TWN work packages were designed around issues identified as important pillars in sustainable wildlife tourism development. Together they were intended to cover a wide area of subjects, building a solid foundation of practical knowledge and scientific fact. Through the project period, it became evident that in clusters where active partnership developed, full potential of the project was truly reached, but in clusters where the partnership never really took off, the success was accordingly.

Some signs during the last phase of the project showed that without supervision from the researchers, the entrepreneurs became more ignorant of the results provided by the researchers. Therefore, the main lesson learned is that for a project like this to truly take off, it is essential to have one or another form of arbitrator to supervise the conduct of an inherent competition or conflict in the process. In this case, the authorities participated as entrepreneurs, not as supervisors or controllers. And a fourth leg is probably needed in the model, the local residents, when tourism is concerned.

References

Arler F (2002) En helt igennem demokratisk natur (A totally democratic nature). In: Agger P (ed) Naturens værdi (The value of nature). Gad, Copenhagen

Ballantyne R, Packer J, Hughes K (2009) Tourists' support for conservation messages and sustainable management practices in wildlife experiences. Tour Manage 30:658–664

Bosetti V, Gerlagh R, Schleicher SP (2009) Modelling sustainable nature – transitions to a sustainable future. Elgar, Cheltenham

Briassoulis H (2002) Sustainable tourism and the question of commons. Ann Tour Res 29(4):1065–1085

Cassini MH (2001) Behavioural response of South American fur seals to approach by tourists – a brief report. Appl Anim Behav Sci 71:341–346

Catlin J, Jones R (2010) Whale shark tourism at Ningaloo Marine Park: a longitudinal study of wildlife tourism. Tour Manage 31:386–394

Curtin S, Richards S, Westcott S (2009) Tourism and grey seals in South Devon: management strategies, voluntary controls and tourists' perceptions of disturbance. Curr Issues Tour 12(1):59–81

Fennell D (2007) Ecotourism, 3rd edn. Routledge, London

Freeman RE (1984) Strategic management: a stakeholder approach. Pitman, Boston

Garrod B, Fennell DA (2004) An analysis of whale watching codes of conduct. Ann Tour Res 31(2):334–352

Granquist SM, Nilsson PÅ (2011) Harmony between man and nature: the case of seal watching at Vatnsnes Peninsula, Iceland. Paper presented at the 3rd international conference on solutions on harmonising sustainability and nature protection with socio-economic stability, at Vidzeme University, Valmiera, Latvia, 19 August 2010

Grant M (2002) PACE: guiding rural tourism development in a fragile area. In: Hall D (ed) Tourism in peripheral areas. Channel View Publications, Clevedon

Gray B (1989) Collaborating. Finding common ground for multiparty problems. Jossey-Bass Publishers, San Francisco

Gössling S, Hall CM (2006) Uncertainties in predicting tourist flows under scenarios of climate change. Clim Chang 79(3):163–173

Hall CM (2008) Tourism planning: policies, processes and relationship. Prentice Hall, London

Hall CM, Lew AA (1998) The geography of sustainable tourism development: an introduction. In: Hall CM, Lew AA (eds) Sustainable tourism: a geographical perspective. Wesley Longman Limited, Harlow, pp 1–12

Hall D, Mitchell M, Roberts L (2003) Tourism and the countryside: dynamic relationship. In: Hall D, Mitchell M, Roberts L (eds) New directions in rural tourism. Ashgate, Aldershot, pp 3–18

Hardy AL, Beeton RJS (2001) Sustainable tourism and maintainable tourism: managing resources for more than average outcome. Ann Tour Res 22(1):186–204

Hughes P (2002) Animals, values and tourism – structural shift in UK dolphin tourism provision. Tour Manage 22:321–329

Hughey K, Ward J, Crawford K, McConnell L, Phillips J, Washbourne R (2004) A classification framework and management approach for the sustainable use of natural assets used for tourism. Tour Manage 25:349–363

Hull J (2001) Opening up the big land to the world. In: Sahlberg B (ed) Going north: peripheral tourism in Canada and Sweden. ETOUR, Ostersund, pp 47–78

Jamal T, Getz D (1995) Collaboration theory and community tourism planning. Ann Tour Res 22(1):186–204

Johnson MW, Lavigne DM (1999) Mass tourism and the Mediterranean Monk Seal – the role of mass tourism in decline and possible future extinction of Europe's most endangered marine mammal, *Monachus monachus*. Monachus Guard 2(2):1–30

Johnston ME (1995) Patterns and issues in Arctic and Subarctic tourism. In: Hall CM, Johnston ME (eds) Polar tourism. Wiley, Chichester, pp 27–42

Kent CSP, Crabtree B (2008) The effectiveness of an established sanctuary zone for reducing human disturbance to Australian Sea Lions (*Neophoca cinerea*) at Carnac Island, Western Australia. Tour Marine Environ 14:29–42

Kovacs KM, Innes S (1990) The impact of tourism on harp seals (*Phoca groenlandica*) in the Gulf of St. Lawrence, Canada. Appl Anim Behav Sci 26(1–2):15–26

Leydesdorff L (2005) The Triple Helix Model and the study of knowledge based innovation. Int J Contemp Sociol 42(1):1–16

Lück M (2007) Nautical tourism development: opportunities and threats. In: Lück M (ed) Nautical tourism: concepts and issues. Cognizant Communication Corporation, New York

Lück M, Higham JES (2008) Marine wildlife and tourism management: scientific approaches to sustainable management. In: Higham JES, Lück M (eds) Marine wildlife and tourism management – insights from the natural and social sciences. Cabi, Wallingford, pp 380–388

Mason P (1997) Tourism codes of conduct in the Arctic and sub Arctic region. J Sustain Tour 5(2):151–165

Nilsson PÅ (2002) Rekreation of tilgængelighed i et tætbefolket område – brug of attityder i det agrare Danmark (Recration and accessibility in a densely populated are – use and attitudes in the agrarian Denmark). Centre for regional and tourism research, Nexø

Nilsson PÅ (2008) Tourism in cold water islands: a matter of contract? Experiences from destination development in the Polar North. Island Stud J 3(1):97–112

Orams MB (2002) Feeding wildlife as a tourism attraction: a review of issues and impacts. Tour Manage 23:281–293

Quiros AL (2007) Tourist compliance to a code of conduct and the resulting effects on whale shark (*Rhincodon typus*) behavior in Donsol, Philippines. Fish Res 84(1):102–108

Ritchie B, Crouch G (2003) The competitive destination: a sustainable tourism perspective. CABI, Wallingford

Rodger K, Moore S, Newsome D (2009) Wildlife tourism, science and actor network theory. Ann Tour Res 36(4):645–666

Strong P, Morris SR (2010) Grey seal (*Halichoerus grypus*) disturbance, ecotourism and the Pembrokeshire Marine Code around Ramsey Island. J Ecotour 9(2):117–132

Timur S, Getz D (2002) Applying stakeholder theory to the implementation of sustainable urban tourism. In: Wöber K (ed) City tourism. Springer, Amsterdam, pp 194–210

Vannini P, Baldacchino G, Guay L, Royle S, Steinberg P (2006) Recontinentalizing Canada: Arctic Ice's liquid modernity and the imagining of a Canadian archipelago. Island Stud J 1(1):3–18

Whoeler EJ, Peney RL, Creet SM, Burton HR (1994) Impacts of human visitors on breeding success and long-term population trends in Adélie Penguins at Casey, Antarctica. Polar Biol 14:269–274

World Commission on Environment and Development (1987) Our common future (Brundtland report) (A/42/427). United Nations, New York

Chapter 9
Tourism Resolving Crisis?

Exploring Tourism Development in Iceland in the Wake of Economic Recession

Gunnar Thór Jóhannesson and Edward H. Huijbens

Abstract This chapter explores the growth of tourism in Iceland in the context of tourism policy and strategy as it manifests in the wake of the financial crash in autumn 2008. Tourism in Iceland has grown from being practically nonexistent in the mid-twentieth century to being one of the three key sectors of the economy. This growth and the role of tourism in the economy has been sporadically recognized with interest in tourism in public debates and policy flaring up in times of crisis in the nation's primary industries. Similarly, after the 2008 credit crunch, great hopes were tied to tourism and a kind of gold rush mentality ensued. The main tenent of the pro-tourism argument is the economic value of tourism and this argument is readorned each time tourism surfaces in public policy. Thus, tourism policy is seen as recurring, thus maintaining the tourism production system, ultimately producing Iceland as a tourism destination. The chapter concludes by arguing that the recurrence of policy in the context of Icelandic tourism fundamentally challenges expectations and hopes tied to tourism in the wake of the economic recession of 2008.

Keywords Tourism • Iceland • Policy • Economy • Lefebvre

G.T. Jóhannesson, Ph.D. (✉)
Institute of Life and Environmental Sciences, Department of Geography and Tourism, University of Iceland, Askja v. Sturlugata, 101 Reykjavík, Iceland
e-mail: gtj@hi.is

E.H. Huijbens, Ph.D.
Icelandic Tourism Research Centre, University of Akureyri, Borgum v/Norðurslóð, 600 Akureyri, Iceland
e-mail: edward@unak.is

D.K. Müller et al. (eds.), *New Issues in Polar Tourism: Communities, Environments, Politics*, DOI 10.1007/978-94-007-5884-1_9,

9.1 Introduction

Iceland has a long history as a destination of travelers and explorers (Ísleifsson 1996). However, tourism as an industry started only to develop in the wake of the Second World War and picked up pace especially after 1980. A key indicator of this growth is departure figures of those showing a foreign passport at Keflavík International Airport, through which 97% of travelers to Iceland have hitherto passed. These figures have multiplied and for instance grew by 66% between 2000 (303,000) and 2010 (502,000) (Ferðamálastofa 2011). This booming expansion of tourism on this island of only 320,000 people has lodged the industry squarely as one of the three pillars of the Icelandic economy, along with fisheries and the aluminum industry. Yet, at current, tourism in Iceland is dealing with several challenges to its industry's development. At the grandest of scales, Icelandic tourism challenges can be framed under the challenges faced by all destinations in the circumpolar region. These include in sociocultural terms the scarcity and homogeneity of people, i.e., small communities often closely knit by blood and kinship; in environmental terms, the fragility of the arctic flora and fauna; and in economic terms, the ubiquity of resource-oriented industries dominating policy and general ethos. Some challenges are common to that of island tourism as framed by (Baldacchino 2006; Lockhart 1997) and Nordic tourism in particular (Hall et al. 2009), such as stubborn patterns of seasonality and high concentration of tourists at few attractions that undeniably put pressure on the fragile natural environment (Jóhannesson et al. 2010). Other current challenges are more particular to Iceland especially and include those posed by the industry structure as being dominated by one major tour operator and the gold rush mentality that ensued in tourism development in the wake of the financial crash that occurred in Iceland in October 2008 (Jóhannesson and Huijbens 2010).

The chapter is intended to place the intensification of tourism demand and mobilities and post crash tourism development rhetoric in the context of tourism policy and strategy as it manifests in the wake of the crash. The chapter will progress in two halves. The former will revolve around theorizing tourism destinations as spatial productions and more particularly as a set of recurring themes and refrains. We seek to clarify our approach through the rhythmanalysis of Henri Lefebvre (2004). For Lefebvre (2004), rhythmanalysis was the ultimate expression of the production of space showing "the interrelation of understandings of space and time in the comprehension of everyday life" (Elden 2004b:vii). He strived to present a nonlinear understanding of time in the form of rhythms and cycles in order to grasp the ever-on-going nature of the everyday. With rhythms Lefebvre (2004) moves away from a reductive account of time seen as a linear progression. For him the linear account of time is an abstraction of time, a conceptualization opposed to the concrete ground of the everyday where Lefebvre saw infinitely superimposed rhythms constituting space, time, and the body. Hence, rhythms are the most concrete of things, the basic underlying organizing principle of everyday life that keeps it in a perpetual state of becoming. Through rhythms destinations emerge through diverse set of repetitive practices. We will focus on those that accumulate to a tourism strategy, which impacts industry and destination development. For us, tourism policy is seen as recurring or a set of refrains which maintains the tourism production system, ultimately producing Iceland as a tourism destination.

The second half of the chapter deals with the central characteristics of Icelandic tourism policy and how it has gone through cycles of reworking in the last decades based on variations of the theme how tourism is to be the salvation of the Icelandic economy. The analytical discussion is based on policy reports on tourism, issued by the government of Iceland; parliamentarian documents that relate to tourism; and interviews with individuals that all have a long and extensive experience in tourism development in Iceland. Among them are former and current ministers of tourism, the former and current general directors of the Icelandic Tourist Board, and the director of the Icelandic Travel Industry Association. The chapter concludes by arguing that the recurrence of policy in the context of Icelandic tourism fundamentally challenges expectations and hopes tied to tourism in the wake of the financial crash of 2008.

9.2 Understanding Recurrent Themes

> Thought is placed in action and action is placed in the world
>
> Anderson and Harrison (2010:11)

The existential truism quoted above represents for us the core of tourism and tourists' being. The growth of tourism in Iceland locally manifests a global intensification of demand, which has spurred a concomitant growth and diversification in the production of tourist destinations and touristic spaces. Britton (1991) frames this relationship of what is perceived and actually produced through what he terms the "tourism production system," drawing on world system dependency theorizing. It refers to the various "commercial and public institutions designed to commodify and provide travel and touristic experiences" (p. 455). Fundamental to us is that in Britton's view, a dynamic interplay between these actors not only creates destinations through attaching social meaning to places but also fundamentally impacts the materiality of the place or the way in which it is ordered (Franklin 2004). Similarly, tourism has been framed in terms of mobilities. Accounting for its hybrid nature, the tourism production system depends on mobilities but also immobilities or infrastructures, local practices, material cultures, and discourses (Sheller and Urry 2004). Hence, tourism as a form of mobility works through a complex topology of overlapping near and far connections and relations that are "produced through practices and relations of different spatial stretch and duration" (Amin 2002:389). These practices and relations are then networks that produce and order places as "material natures, social relations and discursive conceptions" (Bærenholdt et al. 2004:26). This ordering of things and people creating and maintaining place implies that "[t]he only way to understand the durability of orderings (or collections of orderings) [e.g. destinations] is to trace the relations between the heterogenous elements that compose them" (Anderson and Harrison 2010:18). It follows that in order to understand the production of tourist places, it is necessary to trace their interplay in the material setting of a destination.

The above might sound as if the world is a seething mass of malleability, never the same from 1 day to the next. Yet destinations maintain popularity and can build reputations, moving through a product-like life cycle as famously argued by Butler (1980, 2006). What thus needs to be understood is the "... stability of form amid the dynamism of formation" (Anderson and Harrison 2010:19). We would like to follow Crang (2001) as he argues that "[t]hinking of the rhythms of particular locales begins to offer a better grasp on the linking of space and time" (p. 206). We therefore turn to discussion on how destinations can be thought through rhythms and these be analyzed through rhythmanalysis.

9.3 Rhythms of Destinations

Based on the discussion above, it can be stated that destinations are produced in and through socioeconomic relations and practices in a dialectical oscillation of the material (concrete) and the ideal (abstract) (Lefebvre 1991). Within this dialectical oscillation the co-constitutive trajectories of destinations in formation become subject to analysis. For Lefebvre (1991), space itself is born out of the contradictions within the relations of production at the same time it profoundly shapes the apparatus of production. So "productive forces do not merely operate within space but on space, and space equally constrains them" (Elden 2004a:144). Lefebvre (1991) says "space is at once *work* and *product* – a materialisation of 'social being'" (p. 101–102, emphasis original). Elden (2004a:44) summarizes

> There is not the material production of objects and the mental production of ideas. Instead, our mental interaction with the world, our ordering, generalizing, abstracting, and so on produces the world that we encounter, as much as the physical objects we create. This does not simply mean that we produce reality, but that we produce how we perceive reality.

Lefebvre (1991, 2004) proposed to analyze this dialectical oscillation through what he termed rhythmanalysis. He argued that rhythmanalysis was the ultimate expression of the production of space. With it he aimed to show "the interrelation of understandings of space and time in the comprehension of everyday life" (Elden 2004b:vii). Presenting a nonlinear understanding of time in the form of rhythms and cycles, his aim was to grasp the ever-on-going nature of the everyday. He saw rhythm as the most concrete of things, the basic underlying organizing principle of everyday life that keeps it in a perpetual state of motion yet stable.

Rhythm, as formulated by Lefebvre (2004), is a general concept that, although everywhere, eludes meaning. The constitutive part of rhythms or the "first element" of rhythmanalysis that needs to be fleshed out is repetition, i.e., all rhythms entail repetition. But Lefebvre argued that there can be no pure repetition or no absolute, unchanging repetition. An absolute, unchanging rhythm is a pure logical abstraction, a kind of eternal machine, and does not exist in reality. With this argument he stated further that repetition, as it cannot be absolute, gives rise to repetition holding within itself difference, i.e., sooner or later repetition will give rise to difference. Rhythmic cycles of becoming contain always potentialities of change. But although difference is inherent in the repetition and repetition will sooner or later give rise to it, the abstracted repetition also has a role to play. Just as in the production of space where the transcendental conceived (abstract) spaces had real concrete manifestations playing a role in people's everyday life, in the same way implicit with all notion of rhythm, there is measure, i.e., an abstracted ideal that functions as "law, calculated and expected obligation, a project" (Lefebvre 2004:8) that obviously will have their concrete manifestations, although never in accordance with the ideal. The manifest examples of rhythms Lefebvre (2004) cited through his book are of music, language and media, bodily training in terms of dressage and acquired gestures, dance, and work. Although these examples can be understood as abstractions or pure measure, they are all intricately bound with difference since although "rhythm appears as regulated time, governed by rational laws, [it is] in contact with what is least rational in human being: the lived, the carnal, the body" (ibid. p. 9). With these two constitutive elements of rhythm, i.e., as difference and measure, Lefebvre can claim that rhythm has logic although it inherently escapes logic. This interaction of the cyclical (difference producing repetition) and linear (measure) has a folding of interacting component rhythms or polyrhythmia that constitutes the concrete or the present. "In short, things repeat always by virtue of what they are not and do not have" (Deleuze 1994:340).

Here we want to specifically pick up on Lefebvre's formulation of language and media through rhythm analysis:

> Producers of the commodity *information* know empirically how to utilise rhythms. They have cut up time; they have broken it up into *hourly slices*. The output (rhythm) changes according to intention and the hour. (Lefebvre 2004:48, emphasis original)

Rhythms demonstrate to us how all kinds of material and nonmaterial trajectories cutting through space make it alive and malleable yet perceivable as stable and unchanging. This means that figuring a destination is not a matter of calculation or superimposing ideas, images, and rhetoric on an already existing surface, e.g., place or country (Elden 2004a:15–16), but much rather about picking apart the constitutive threads of destination making as they recur in different guises. As a result a destination is only seemingly stable and incessantly ripe with potential to change "according to the intention of the hour."

The methodological approach in this chapter is qualitative. Data was gathered through semi-structured interviews with practitioners and policy makers in tourism development in Iceland and officials working in the public administration, including four out of six former ministers of tourism since 1989. The focus was on the general trends in tourism policy in Iceland. The interviews have been conducted from January 2009 to April 2011. The following discussion also rests on analysis of various policy documents issued by public authorities and public discourse on tourism development as manifested in the two of the largest newspapers in Iceland and public speeches by the minister of industry and tourism.

9.4 Icelandic Economy and the Tourism System

During the last century, Iceland has gone through accelerated modernization. After having been among the poorest nations in Europe in the late nineteenth century (Ólafsson 2005), the country topped the United Nations Human Development Index in 2008 (UNDP 2008). Economic monoculture is common to arctic and subarctic regions and the prevalence of resource-oriented industries. In Iceland, the national economy has predominantly been based on fisheries, leaving it open to the vagaries of the open market that has resulted in immense economic fluctuations (Ólafsson 2008). During the second half of the twentieth century, systematic attempts of economic diversification were made with special focus on harnessing the rich resources of geothermal energy and hydropower, translated into an emphasis on aluminum production which together with the growing service sector has led to decreasing dependence on the fisheries. Arguably a decisive shift has occurred from a resource-based economy toward service-based economy (Jóhannesson et al. 2003). Nonetheless, fisheries were in 2009 still the largest source of foreign currency receipts (29%) with aluminum production second (24%) and tourism in third place (13%) (Statistics Iceland 2011), although taking into account the balance on current accounts, the prospects for tourism seem greater than aluminum, although it remains to be researched. After privatization of the largest banks in Iceland, finalized in 2003, the financial sector boomed and was by many regarded as the fourth pillar of the economy (Gissurarson 2002; Viðskiptaráð Íslands 2006). In 2000 the total size of the three largest banks in Iceland amounted to just over 100% of the GNP of the country. In the end of 2007, their size was nine times that of the annual GNP (Hreinsson et al. 2010). When the global financial crises ensued, the three largest, highly leveraged, Icelandic banks were among the first to suffer.

In autumn 2008 the Icelandic financial sector crumbled with more than 80% of the banking sector collapsing in matter of a few days (Matthiasson 2008). The national currency plummeted, inflation rose steeply, and soon the unemployment rate increased dramatically (stands in 8.1% in April 2011) (Vinnumálastofnun 2011). Now, 3 years on, the economic challenges are being sorted out, but Table 9.1 compares the state of the Icelandic economy before and after the financial meltdown.

As the table demonstrates, "the response of the real economy has been significantly milder than feared at the onset of the crisis" (Halldórsson and Zöega 2010:31–32). The stock of real assets is great, public sector debt is similar to around the year 2000, and the islanders remain on average among the richest nations. Bulk of the losses during the meltdown ended up on foreign creditors. This is not to say that the meltdown had no negative impact. Many have suffered greatly both households and companies domestically, mostly those that were very heavily indebted at the time of the crash (Magnússon 2010). What the crash seems to have provided is a reality check in a sense, refocusing public attention toward the sectors of the economy which generated foreign revenues from direct sales of goods and/or services instead of speculated profits. As we have argued elsewhere, tourism was one of those sectors (Jóhannesson and Huijbens 2010).

Table 9.1 Macroeconomic indicators for Iceland, comparison 2003–2007 and 2008–2009 (annual average growth (%) unless indicated otherwise) (Source: Halldórsson and Zöega 2010:6)

Macroeconomic indicators for Iceland (annual average growth (%) unless indicated otherwise)		
National accounts	2003–2007	2008–2009
Real GDP growth	5.6	−2.8
Real private consumption growth	7.0	−11.3
Real gross capital formation growth	17.2	−35.5
Gross capital formation (% of GDP)	26.9	14.1
Real exports growth	6.2	6.7
Real imports growth	12.8	−21.1
Current account (% of GDP)	−14.3	−3.6
Savings		
Gross household saving (% of disposable income)	16.3	23.9
Public saving (% of GDP)	2.8	−9.1
Money and banking sector		
Real money stock (M1) growth	41.9	−0.1
Real domestic lending of banking sector growth	29.9	−34.2
Asset prices		
Real share prices growth	34.6	−72.4
Real residential house prices growth	11.9	−13.4
Labor market		
Unemployment (% of labor force)	2.9	7.2
Unemployment growth (% of itself)	−6.2	85.2
Unemployment growth (2009, % of itself)	–	140.0
Total employment growth	2.5	−2.7
Total employment growth (% of pop.)	82.0	80.9
Average real wage growth	2.8	−5.5
Private sector real hourly earnings growth	3.2	−6.3
Labor share (% of GDP)	68.5	56.8
Other variables		
Business bankruptcies (yearly average number)	598	910

9.5 The Tourism System in Iceland

Despite tourism becoming in real terms a prominent figure in Icelandic national accounting, it has only sporadically surfaced as a development option and has yet to be promoted as such in any coherent manner (Jóhannesson et al. 2010). The Icelandic government has mostly remained passive toward tourism development that has in turn meant that private stakeholders have, for better or worse, had much freedom in embarking on their own tourism projects.

Tourism development in Iceland is characterized by this entrepreneurial freedom but at the same time a heavily skewed industry composition. It is dominated by few large firms and numerous small to micro-size, family-run businesses that are to a large extent reliant on the marketing strategies and vision of the dominant players.

Tourism in Iceland is heavily dependent on transport by air and the infrastructure built by the US and British forces during Second World War. At that time the geographic location of the island made it ideal for the development of air transport, a fact utilized by Icelandic airlines to establish extensive route networks both to mainland Europe and North America soon after the Second World War. The geographical location of Iceland provided a natural hub in the middle of the North Atlantic which promoted stopovers between Europe and North America. The fact that Icelandic airlines have been able to strengthen the route network connecting the island to major markets is a central precondition of tourism development in Iceland, as it has allowed access to a much larger pools of potential passengers flying between the continents. The airlines have sufficiently rationalized their operations and are able to sustain the hitherto almost remarkable growth in visitor arrivals. Today two Icelandic airlines operate flights to 53 destinations in the summer of 2011, with several more carriers now offering flights to the island during the summer months. Icelandair, which builds on the postwar history of Icelandic cross-continental aviation, is part of by far the largest tourist firm in the country. This firm is Icelandair group that, in addition to the international carrier Icelandair, includes the largest domestic airline, a hotel chain, and tour operator, which makes it truly the ruling actor in Icelandic tourism controlling almost half of the domestic supply, as calculated from the proportion of inbound tourists they service and revenue generated from these (Icelandair Group 2011; Statistics Iceland 2010).

With this dominance and disparity within the private sector, the public sector has not played a great role in tourism development. The industry has had hard times in gaining recognition by the authorities. The government has though made repeated attempts to devise holistic tourism policies and agendas on particular issues (Samgönguráðuneytið 2001, 2003). The industry has however perceived these attempts as halfhearted since support to the tourism sector has always been minimal, with little consideration to its importance in the national economy. This has meant, in the words of Kjartan Lárusson, the former chairman of the Icelandic Tourist Board, that "the sector has been disconnected from the public policy but has kept going [. . .]" (interview quote, Kjartan Lárusson, April 28 2009).

9.6 Toward the Rhythms of Tourism Policy

> Generally flotsam, with luck a monster: an order, a prayer
>
> Lefebvre (2004:46)

In the wake of the financial collapse, tourism appeared in the limelight of public discourse on how to resolve the crisis. This is not the first time tourism appears as an option for economic regeneration during times of recession (Jóhannesson and Huijbens 2010). The history of policy making for tourism manifests a cyclical trend. As such, the notion of tourism development seems to sporadically attract political attention in tandem with the waxing and waning of the primary industries. In order to tease out how this particular rhythm of engagement with tourism challenges sustainable tourism development, we will trace how it has played out in the past and identify its recurrence in the present.

In the following the focus is on the period between 1990 and 1996 when the first Icelandic tourism policy was formalized by the ministry of communication, transport and tourism (Samgönguráðuneytið 1996). This event followed a series of unsuccessful attempts to create a tourism policy, most notable in 1973, 1983, and 1990 (Checchi & co. 1975; Samgönguráðuneytið 1983, 1990). All of these documents address similar themes: planning, environment and sustainability, image and marketing, and research and education as did the one finally produced and published in 1996. In outlining this transition from failure to success, our story mostly takes place within the national parliament of Iceland; Alþingi.

In December 1990 a bill for a new tourism policy was proposed by the leftist government at the time based on the work by a cross political committee (Alþingi 1990). Steingrímur J. Sigfússon, the minister of communication, transport and tourism, was its proposer. The bill was rather general in terms, albeit concerned with the themes identified above. A central objective was to "develop and recognize tourism as important and profitable industry" (Alþingi 1990:1). More particular aims to be achieved in order to realize this central objective was related to natural and cultural protection, diversification of the economy, regional development, and improving general living standards. The proposed bill was met with mixed feelings, with main skeptics residing in the right of center Independence Party (Sjálfstæðisflokkurinn). One of its MPs, Halldór Blöndal, opposed the bill strongly on the grounds it was unclear both in its aims and implications. The bill was taken up a few months later (19th of March 1991) by the standing committee of communication and transport for a quick discussion that resulted in a recommendation for a vote. However, thwarted by the opposition, the bill never did go that far. The general director of the Icelandic Tourist Board (1990–1991 and 1993–2007), Magnús Oddsson, recalls this point:

> I was asked to come down to the parliament. I just remember, it was noon and then it is being discussed, this is one of the cases, if it should go through or not. Because the parliament is coming to an end and it is being discussed which cases should go through. And it was perfectly clear there, I just felt it on this meeting I attended that the MPs did not see this as a case to finish. [...] So this was one of the things that were pushed aside and not voted on. (interview quote, Magnús Oddsson January 23 2009)

Soon after this, parliamentary elections were held and a new government was formed on the 30th of April 1991, in which Halldór Blöndal was appointed minister of tourism (as well as minister of communication and transport and agriculture).

Not surprisingly this new minister did not push for the adoption of the tourism policy proposed by the former government, which he was instrumental in thwarting. However, in autumn 1995, the former minister, Steingrímur J. Sigfússon, who was by now in opposition, requested information from the minister of tourism about work on the tourism policy. In his answer, the minister, Halldór Blöndal, mocked the opposition for not having been able to get a tourism policy through during their time in government 4 years earlier but also informed the parliament that a policy paper was indeed being prepared (Alþingi 1995). According to Halldór Blöndal, the main rationale for stressing tourism was the economic situation in the country in the early 1990s. Foreign currency was desperately needed, living standards were falling, and tourism was seen as a potential mean for boosting the economy.

Political mud wrestling aside, what had changed in the meantime is that Magnús Oddsson had readorned the economic rationale for tourism development this time firmly placing it in the context of the production economy. He argued in a column in Morgunblaðið, the most popular newspaper in the country at the time, that the foreign currency revenue generated by each tourist equaled the revenues from one ton of cod, the main export of Icelandic fisheries (Oddsson 1993). This comparison was an eye opener, as Magnús Oddsson recalls:

> ... all of a sudden ... one afternoon, this just became the hot stuff in the debate. All of a sudden this was put in a context that people understood. [...] [...] this was the only way to get people to realize [the economic significance of tourism]. To try to connect it to something ... because people understood cod. (interview quote, Magnús Oddsson January 23 2009)

Subsequently, Magnús Oddsson referred repeatedly to the value of tourism in terms of fish and fisheries, tourists were "catch" and the planes of Icelandair "trawlers." He also made similar comparison to aluminum smelters (e.g., Oddsson 1998) with the aim to frame tourism as a stable and solid provider of foreign currency that should be treated equally by the authorities.

The story of the bill as it was stalled in parliament and latter to be re-proposed by the former opposition illustrates a process of reiteration which has by now been translated into two subsequent tourism policies, the latter still waiting for approval by the parliament (Samgönguráðuneytið 2005; Alþingi 2011). The fact that the first proposed bill was not supported can largely be explained by how tourism had not entered into popular mind-sets and was at the time flotsam in political rhetoric, playing a role in political mud fights, "according to the intention of the hour." The reason for the initial bill to be abandoned was perhaps not so much due to its content or organizational reforms as those had to be negotiated across political parties, as it was the need felt by the right of center opposition party Sjálfstæðisflokkur, to make life as difficult as possible for the left of center government during its last days in power before general elections. Once in government the Sjálfstæðisflokkur (1995), building on the successful comparative rhetoric as introduced by Magnús Oddsson, started its own round of policy discussions many of which concluded with the same considerations as those expressed in 1990.

The 1996 policy created space for a more guided effort in tourism development in Iceland, primarily manifested in comprehensive carrying capacity studies at some of the island's key destinations (Sæþórsdóttir et al. 2001; Aradóttir et al. 2003; Sæþórsdóttir et al. 2003). Furthermore, between 2000 and 2005, numerous reports were produced by the ministry of tourism in what seems like an attempt by the government to engage in the sector (Samgönguráðuneytið 2001, 2003, 2004). However, these reports did not translate into any systematic action in terms of financing basic elements of tourism development, such as research and infrastructure. Tourism withdrew from the limelight in the first decade of the twenty-first century only to resurface in the immediate aftermath of the financial crash in

October 2008. Again the founding argument was that tourism was a tool to save the wounded Icelandic economy. This recurring theme was expressed by the minister of tourism (Skarphéðinsson 2008) and the key stakeholders of tourism, such as the general director of the Icelandic Tourist Board (Atladóttir 2008) and the director of the Icelandic Travel Industry Association (Hauksdóttir 2009). In a similar way as in the 1990s, tourism was tied to the production economies. Consequently, the emphasis was placed on raising the numbers of visitors (Skarphéðinsson 2008). Thus, the mentality in regard to tourism development by the central authorities has to a large extent been similar to the production industries where more fish means more money and larger aluminum smelters mean greater profits.

The rationale of production economy that drives discourse on tourism development at current is not surprising in a country that desperately needs economic revival. After all, the sector has for long waited for the recognition of tourism as a "real" industry, so for them it may sound as a dream coming true. However, underlying this and recurring in all policy documents since the first in 1973 is the need for infrastructure development. Indeed, the Icelandic Tourist Board has regularly issued concerns over lacking infrastructure, which will not be able to sustain too fast growth in number of tourists, but have till now not received backing to do anything about it. There is in other words a wide gap between the policy and practice by authorities that during times of gold rush, mentality in the tourism sector seriously threatens the basic resources of the sector.

Here again comparison can be made with fisheries which are tightly regulated in recognition of the fragility of the resource, while the country's nature and wilderness areas, its key tourism resource, has hardly been made sense of in terms of policy by tourism authorities (government and key stakeholders). Minimum support (100 million ISK) was provided for infrastructure development in 2009 (Iðnaðarráðuneytið 2009; Skarphéðinsson 2009). This lack of policy is manifested in the ministry of environment putting 42 million ISK into emergency infrastructure development at some of Iceland's key destinations in spring 2011 (Umhverfisráðuneytið 2011). The amount is ridiculous in the face of the need but actually huge in terms of the history of tourism infrastructure development in Iceland.

In a recent survey among managers in Icelandic firms, tourism together with the energy sector was seen as offering the best opportunities for growth in the near future (Viðskiptaráð Íslands 2011). They are probably right. This year, 600,000 tourist arrivals are expected to the island and the outlook is that tourist arrivals will pass the one million mark within 10 years, at least that is the aim of the CEO of the leading Icelandic tourism firm, Icelandair Group (Vísir 2011). The question remains, however, if the tourism resources of the island can sustain such a growth. With the Icelandic Tourist Board and several public stakeholders trying to ride the crest of public recognition of tourism in the wake of the credit crunch, the recurring economic rationale that facilitates this recognition seems also to have in store a major detrimental transformation of Iceland's key destinations.

9.7 Conclusion

This chapter has argued for an understanding of destinations and their inherent complexities through the approach of rhythmanalysis (Lefebvre 1991, 2004). From this view, tourism destinations emerge through diverse recurrent themes and enactments. The discussion has focused on one particular set of recurrences, those that occur in language and media in the form of tourism policy in Iceland. As noted above, rhythmic cycles contain potentialities of change and things tend to repeat by virtue of what they are not or what they lack. There are some signs that repetitive enactments of Icelandic tourism policy are moving toward an emphasis put on infrastructure development and systematic approach to planning, put forward in a revised tourism policy (Alþingi 2011), but the rhetoric remains that the aim is increasing the economic yield of tourists. With this recurrent theme, the sector seems to have cemented its position as a development option, both by public and private actors, at the same time hampering efforts of infrastructure development.

Above we have argued how the recurring founding tenent of Icelandic tourism policy fundamentally challenges expectations and hopes tied to tourism in the wake of the financial crash of 2008. Indeed, it seems as if the two constitutive elements of rhythm, i.e., as difference and measure, have both sustained tourism policy initiatives and yet propel it on new trajectories, most likely detrimental to the development of tourism destinations in Iceland.

In order to avoid disastrous effects on the tourism resources of the island, the tourism policy necessarily has to reach stakeholders. The recurrent themes in tourism policy, readorned as they may be according to the intention of the hour, need to reach a wider audience. The authorities need to start living up to their role as a ruling body in tourism development on the island and grab a firm hold of the reigns. At current there is a gap between public authorities and private actors, which has to be bridged. Words on infrastructure development, systematic planning, and quality over quantity have to be translated into practice. This challenge remains unresolved. What this means in terms of the greater challenges faced by polar region in general is that due to the scarcity and homogeneity of people, the fragility of the arctic flora and fauna, and the ubiquity of resource-oriented industries dominating policy and general ethos, public authorities need to play an active role as setting the scene for tourism development to unfold in the polar regions. In Iceland in particular, this means infrastructure development and a move away from resource extraction-oriented policy jargon.

Acknowledgments We would like to thank the editors for their efforts in making this book a reality and for accepting our abstract. We are also grateful to all our interviewees for sharing their views on the development of tourism in Iceland and giving us access to their personal archives of material relating to tourism discourse. The chapter is based on an ongoing research on the development of tourism in Iceland that is funded by the Icelandic Research Fund.

References

Alþingi (1990) Frumvarp til laga um ferðaþjónustu. Þskj. 363, 246. mál (Parliamentary proposal of a bill on tourism). Retrieved 6 Ap 2011, from, http://www.althingi.is/dba-bin/ferill.pl?ltg=113&mnr=246

Alþingi (1995) 18. október 1995: Málefni ferðaþjónustu (Issues of tourism). Retrieved 29 June 2011, from, http://www.althingi.is/dbabin/prentaups.pl?lnr=%2Fwwwtext%2Fhtml%2F120%2F10%2Fl18133632.sgml&pdf=PDF

Alþingi (2011) Tillaga til þingsályktunar um ferðamálaáætlun 2011–2020. Þskj. 758 – 467. mál (Parliamentary resolution on tourism policy). Retrieved 20 Mar 2011, from, http://www.althingi.is/dba-bin/ferill.pl?ltg=139&mnr=467

Amin A (2002) Spatialities of globalisation. Environ Plan A 34(3):385–399

Anderson B, Harrison P (2010) Taking-place: non-representational theories and geography. Ashgate, Aldershot

Aradóttir B, Sæþórsdóttir AD, Gísladóttir G, Ólafsson AM (2003) Þolmörk ferðamennsku í Landmannalaugum (Tourism carrying capacity in Landmannalaugar). Icelandic Tourism Research Centre, The Icelandic Tourist Board, The University of Iceland and the University of Akureyri, Akureyri

Atladóttir ÓÝ (2008) Tækifærin í ferðaþjónustunni (Opportunities in tourism). Morgunblaðið

Baldacchino G (ed) (2006) Extreme tourism: lessons from the world's cold water islands. Elsevier, Oxford

Britton S (1991) Tourism, capital, and place: towards a critical geography of tourism. Environ Plan D Soci Space 9:451–478

Butler RW (1980) The concept of tourist area cycle of evolution: implications for management of resources. Can Geogr 24(1):5–12

Butler RW (ed) (2006) The tourism area life cycle. Vol 1. Applications and modifications. Channel View Publications, Clevedon

Bærenholdt JO, Haldrup M, Larsen J, Urry J (2004) Performing tourist places. Ashgate, Aldershot

Checchi & co (1975) Tourism in Iceland: phase 2. Checchi & co, Washington

Crang M (2001) Rhythms of the City. Temporalised space and motion. In: May J, Thrift N (eds) Timespace: geographies of temporality. London, Routledge, pp 187–207

Deleuze G (1994) Difference and repetition. Continuum, London

Elden S (2004a) Understanding Henri Lefebvre. Continuum, London

Elden S (2004b) Rhythmanalysis: an introduction. In: Lefebvre H (ed) Rhythmanalysis: space, time and everyday life. Continuum, London

Ferðamálastofa (2011) Tourism in Iceland in figures. The Icelandic Tourism Board, Reykjavík

Franklin A (2004) Tourism as an ordering. Towards a new ontology of tourism. Tour Stud 4(3):277–301

Gissurarson HH (2002) Hvernig Ísland getur orðið ríkasta land í heimi (How Iceland can become the world richest country). Nýja bókafélagið, Reykjavík

Hall CM, Müller DK, Saarinen J (2009) Nordic tourism. Issues and cases. Channel View Publications, Bristol

Halldórsson ÓG, Zöega G (2010) Iceland's financial crisis in an international perspective. Economic Institute working paper series W10:02. University of Iceland Economic Institute, Reykjavík

Hauksdóttir E (2009) Ferðaþjónustan í upphafi árs (Tourism at the start of the year). Fréttablaðið

Hreinsson P, Benediktsdóttir S, Gunnarsson T (eds) (2010) Aðdragandi og orsakir falls íslensku bankanna 2008 og tengdir atburðir. Rannsóknarnefnd Alþingis, Reykjavík

Icelandair Group (2011) Investor reports and presentations. Retrieved 25 June 2011, from, http://www.icelandairgroup.com/investors/reports-and-presentations/

Iðnaðarráðuneytið (2009) Ferðamálaþing 2008. (The annual Icelandic tourism conference 2008). Retrieved 18 May 2009, from, http://www.idnadarraduneyti.is/radherra/raedur-greinarOS/nr/2661

Ísleifsson S (1996) Ísland framandi land (Iceland a foreign place). Mál og menning, Reykjavík
Jóhannesson GÞ, Huijbens E (2010) Tourism in times of crisis: exploring the discourse of tourism development in Iceland. Curr Issues Tour 13(5):419–434
Jóhannesson G, Skaptadóttir UD, Benediktsson K (2003) Coping with social capital? The cultural economy of tourism in the North. Sociologica Ruralis 43(1):3–16
Jóhannesson GÞ, Huijbens E, Sharpley R (2010) Icelandic tourism: opportunities and threats. Tour Geogr 12(2):278–301
Lefebvre H (1991) The production of space. Blackwell, Oxford
Lefebvre H (2004) Rhythmanalysis. Space, time and everyday life. Continuum, London
Lockhart D (1997) Islands and tourism: an overview. In: Lockhart D, Drakakis-Smith D (eds) Island tourism: trends and prospects. Pinter, London, pp 3–20
Magnússon G (2010) Lessons from a small country in a financial crisis or Dr. Minsky and Mr. Ponzi in Iceland. Economic Institute working paper series W10:03. University of Iceland Economic Institute, Reykjavík
Matthiasson T (2008) Spinning out of control, Iceland in crisis. Nord J Pol Econ 34(3):1–19
Oddsson M (1993) Gjaldeyristekjur af hverjum erlendum ferðamanni nær þær sömu og af hverju tonni af þorski (Foreign currency income by each tourist almost the same as by one ton of cod). Morgunblaðið
Oddsson M (1998) Innfluttir Skattgreiðendur og neytendur: Grein í Morgunblaðinu janúar 1998 (Imported tax payers and consumers: article in Morgunblaðið January 1998). Retrieved 30 June 2011, from, http://ferdamalastofa.is/Category.mvc/Display/270#Grein%20%C3%AD%20Morgunbla%C3%B0inu%20jan%C3%BAar%201998
Ólafsson S (2005) Normative foundations of the Icelandic welfare state: on the gradual erosion of citizenship-based welfare rights. In: Kildal N, Kuhnle S (eds) Normative foundations of the welfare state: the Nordic experience. Routledge, London/New York
Ólafsson S (2008) Íslenska efnahagsundrið: Frá hagsæld til frjálshyggju og fjármálahruns (The Icelandic economic wonder: from prosperity to neo-liberalism and financial crash). Stjórnmál og stjórnsýsla veftímarit 4(2):231–256
Sæþórsdóttir AD, Gísladóttir G, Ólafsson AM, Sigurjónsson BM, Aradóttir B (2001) Þolmörk ferðamennsku í þjóðgarðinum í Skaftafelli (Tourism carrying capacity in Skaftafell National Park). Icelandic Tourism Research Centre, The Icelandic Tourist Board, The University of Iceland and the University of Akureyri, Akureyri
Sæþórsdóttir AD, Gísladóttir G, Aradóttir B, Ólafsson AM, Ólafsdóttir G (2003) Þolmörk ferðamennsku í friðlandi á Lónsöræfum (Tourism carrying capacity in Lónsöræfi). Icelandic Tourism Research Centre, The Icelandic Tourist Board, The University of Iceland and the University of Akureyri, Akureyri
Samgönguráðuneytið (1983) *Ferðamál á Íslandi: Þjóðhagsleg þýðing, spá um þróun til 1992, tillögur um ferðamálastefnu* (Tourism in Iceland: national economic weight, prediction till 1992 and proposals for a strategic plan). Samgönguráðuneytið, Reykjavík
Samgönguráðuneytið (1990) Ferðamálastefna: Áfangaskýrsla Ferðamálanefndar samgönguráðuneytisins (Tourism policy: an interim report of the tourism committee to the ministry of transport). Samgönguráðuneytið, Reykjavík
Samgönguráðuneytið (1996) Stefnumótun í ferðaþjónustu (Tourism strategy). Samgönguráðuneytið, Reykjavík
Samgönguráðuneytið (2001) Menningartengd ferðaþjónusta (Cultural tourism). Samgönguráðuneytið, Reykjavík
Samgönguráðuneytið (2003) Íslensk ferðaþjónusta: Framtíðarsýn (Icelandic tourism: future prospects). Samgönguráðuneytið, Reykjavík
Samgönguráðuneytið (2004) Afþreying í ferðaþjónustu: Skýrsla nefndar samgönguráðherra um öryggisreglur og leyfisveitingar í afþreyingarferðaþjónustu (Entertainment in tourism: report on safety rules and licence provisions in adventure tourism). Samgönguráðuneytið, Reykjavík
Samgönguráðuneytið (2005) Ferðamálaáætlun 2006–2015 (Tourism policy). Samgönguráðuneytið, Reykjavík

Sheller M, Urry J (eds) (2004) Tourism mobilities: places to play, places in play. Routledge, London/New York

Skarphéðinsson Ö (2008) Ræða ráðherra á ferðamálaþingi 2008 (Minister's speech at the annual Icelandic tourism conference 2008). Retrieved 22 Jan 2009, from, http://idnnadarraduneyti.is/radherra/raedur-greinar/nr/2661

Skarphéðinsson Ö (2009) Brúarsmíði sjóða og sprota (Bridgebuilding of funds and innovators). Morgunblaðið

Statistics Iceland (2010) Tourism satellite accounts 2000–2008. Hagstofa Íslands, Reykjavík

Statistics Iceland (2011) Útflutningur vöru og þjónustu eftir atvinnugreinum 1990–2009 (Export of goods and services by industry). Retrieved 8 June 2011, from, Hagstofa Íslands: http://www.hagstofa.is/temp/Dialog/Saveshow.asp

Umhverfisráðuneytið (2011) Varið verður 42 milljónum til brýnna framkvæmda á friðlýstum svæðum nú í vor (42 million will be spent on urgent infrastructure development in protected areas this spring). Retrieved 29 June 2011, from, http://www.umhverfisraduneyti.is/frettir/nr/1794

UNDP (2008) The human development index – going beyond income. Retrieved 15 Apr 2009, from, http://hdrstats.undp.org/2008/countries/country_fact_sheets/cty_fs_ISL.html

Vinnumálastofnun (2011) Staða á vinnumarkaði: Apríl 2011 (Employment situation: April 2011). Retrieved 8 June 2011, from, http://www.vinnumalastofnun.is/files/apr%C3%ADl%2011.pdf

Viðskiptaráð Íslands (2006) Ísland 2015 (Iceland 2015). Icelandic Chamber of Commerce, Reykjavík

Viðskiptaráð Íslands (2011) Tökumst á við tækifærin: atvinnulíf til athafna (Let's grapple with the opportunities: industries to action). Viðskiptaráð Íslands, Reykjavík

Vísir (2011) Stefna að því að fá milljón ferðamenn til landsins (Aim to have one million visitors to the country) Retrieved 14 June 2011, from, http://www.visir.is/stefna-ad-thvi-ad-fa-milljon-ferdamenn-til-landsins-/article/2011110318871

Chapter 10
A (Ski) Trip into the Future

Climate Change and Winter Tourism in Polar Sweden in 2040

Patrick Brouder and Linda Lundmark

Abstract Recent research shows that many polar tourism entrepreneurs are not ready to adapt to climate change even though some areas within the polar north are more exposed and adaptive capacities of entrepreneurs vary. Moreover, sensitivities of communities to climate change are dependent on the importance of current and future winter tourism in the locale relative to other industries, thus questioning whether communities should focus on winter-tourism development. Current trends reveal a promulgation of a four seasons' approach to tourism in Polar Sweden in order to reduce sensitivities. However, there is also further winter-tourism development potential since the north inland of Sweden is more secure than many other European winter sport destinations. An important question is as follows: how can regions in the polar north capitalize on their natural assets and develop winter tourism without risking a lock-in effect through increased regional sensitivity to climate change? This chapter maps the exposed area in Polar Sweden and generates basic climate impact scenarios for the future based on plausible alternatives due to climate change and tourism business and institutional development. The discussion utilizes the Arctic Tourism Innovation System (ATIS) framework, where the necessary roles and complementary relationships of institutions and entrepreneurs in creating sustainable paths for polar communities are highlighted.

Keywords Arctic tourism innovation system • Climate change • Entrepreneurship • Institutions • Sweden

P. Brouder • L. Lundmark, Ph.D. (✉)
Department of Geography and Economic History, Umeå University, SE-90187 Umeå, Sweden
e-mail: patrick.brouder@geography.umu.se; linda.lundmark@geography.umu.se

D.K. Müller et al. (eds.), *New Issues in Polar Tourism: Communities, Environments, Politics*, DOI 10.1007/978-94-007-5884-1_10,

10.1 Climate Change and Winter Tourism in Polar Sweden

Climate change in the polar north is both an opportunity and a challenge for communities. The Arctic Climate Impact Assessment states: "mean annual Arctic surface temperatures north of 60° N will be 2–4°C higher by mid-century… Precipitation is projected to increase by about 8% by mid-century" (ACIA 2004, p. 994). If the increasing temperature and precipitation result in more pleasant conditions for outdoor recreation, then the future prospects for tourism in Polar Sweden are bright. If, however, they result in warmer winters with more frequent freeze-thaw cycles and precipitation in the form of winter rain, then the future prospects for winter tourism in particular are questionable. This chapter looks at changes in winter-oriented tourism activities in the year 2040 since this is the most vulnerable type of tourism in the context of climate change although it is acknowledged that, relative to other European regions, Polar Sweden is more secure within this time frame.

The aim of the chapter is to discuss what winter tourism may look like in Polar Sweden in 2040. The polar area in this study includes the counties of Norrbotten and Västerbotten (cf. Lundmark 2010). The purpose is to allow a more considered discussion on climate change impacts on the winter-tourism sector within a time frame beyond individual firms' horizons yet still within the planning perspective of regional authorities. To this end, the elements of vulnerability to climate change – exposure, sensitivity, and, particularly, adaptive capacity – are examined from an intra-regional perspective within the study area, and scenarios are generated for a number of possible futures.

It is important to highlight from the outset that vulnerability here is a relative concept since, within the proposed time frame, according to available climate change scenarios (Lind and Kjellström 2008), the region is not so exposed to climate change that winter tourism will end nor is any locale within the region too sensitive since tourism never represents more than 10% of employment in any municipality annually; however, it varies seasonally (Lundmark 2005), and adaptation is, in the long term, a possibility for all businesses. Moreover, natural asset capitalization through major investments is an attractive option for Polar Sweden too – new and existing winter sport resorts may be developed if the relative snow security of Polar Sweden begins to attract tourists from all over Europe. This raises the issue of how much and what type of development of tourism, and winter tourism, in particular, is desirable and sustainable from a regional development perspective in the long term. The question is as follows: how much winter-tourism development is feasible in order to maximize the positive economic benefits while avoiding a lock-in effect which may leave some communities more vulnerable in the future? The tourism development paths of communities are emerging and decisions made today by planners have long-term effects which, in combination with entrepreneurial developments, may change the vulnerability equation for certain communities.

10.2 The Local Geographical Context

The research area in this study is the Upper Norrland region of Northern Sweden, made up of the two most northerly and largest counties in the country – Norrbotten and Västerbotten (see Fig. 10.1). The coastal municipalities form the economic center of the region with the inland municipalities having been economically dependent on the primary sector in the past but increasingly being seen as a Swedish pleasure periphery (Müller 2011; Müller and Jansson 2007). The tourism sector is mostly made up of micro-firms (less than ten employees) with the main tourism offer being nature-based tourism. Tourism employment, with notable seasonal variation, may be up to 10% of the workforce in certain municipalities (Lundmark 2005). The area under study lies above 60° N and includes parts of Sápmi – the traditional land of the Sámi people of Northern Europe. Many areas within the region are not

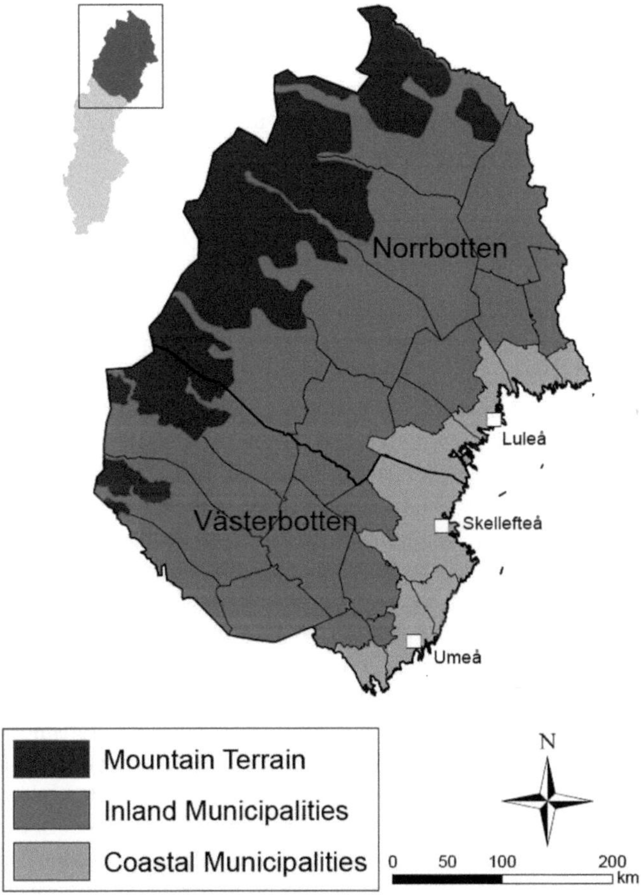

Fig. 10.1 Map of Polar Sweden (Source: Brouder and Lundmark 2011)

expected to be dramatically affected by changes in climatic conditions by mid-century (Lind and Kjellström 2008). However, the Swedish Meteorological and Hydrological Institute expects 20–30 fewer days of snow cover in the coastal region (SMHI 2010). The current supply of tourism firms is thus not drastically exposed since the majority of winter-oriented firms are located in the inland region (Brouder and Lundmark 2011), but there remains a question mark over how planners and entrepreneurs in the coastal areas should develop their winter offer going forward. The inland winter-oriented businesses are not without concern either since future extremes and freeze-thaw cycles may mean they too will have to adapt to climate change in order to remain viable. Polar Sweden differs from many other polar regions yet it remains at the vanguard of the polar pleasure periphery since, while being both relatively more accessible and developed in terms of tourism, it shows how tourism is part of a diversified economy and is thus an interesting case of the potential long-term development paths for polar communities in general, many of which are striving to diversify through (among other means) tourism (Noakes and Johnston 2009).

10.3 Climate Change, Winter Tourism, and Entrepreneurship

10.3.1 Intra-regional Analysis of Climate Change

Community resilience to climate change is set to be a major challenge in the coming decades. Resilience is here defined as the ability of a community to absorb strain placed upon it by exogenous forces. Some communities succeed in adapting to external challenges and some do not, and regarding climate change and tourism, it is clear that there will be winners and losers (Smith 1990), but there is much evidence to suggest that the winners and losers will be located much closer in space than a rudimentary analysis would suggest, and they may even be present within the same region (Brouder and Lundmark 2011; Tervo-Kankare and Saarinen 2011). This intra-regional lens demands an enhanced vulnerability analysis, one that fully explores the exposure, sensitivity, and adaptive capacity (Schneider et al. 2001) of all tourism actors within a given region. Within tourism, analysis of resilience is complicated by the fact that tourism is a complex adaptive system (Becken and Hay 2007), and adaptive actions which are part of the everyday struggle for survival are not easily discernible from specific climate change adaptations nor are adaptive actions limited to tourism matters. In particular, the dominance of micro-firms in tourism means that the nature of entrepreneurship in a given region may affect how effectively and how soon autonomous adaptations take place. In Polar Sweden, there is a lack of immediacy in general to act on what are perceived as small and slow changes due to climate change (as opposed to reactive responses to increasing extreme weather events) even though there are clearly differing vulnerabilities within the region depending on location and product offered.

10.3.2 Vulnerability to Climate Change

Not all municipalities in the polar north will be affected equally by future changes in the regional climate (O'Brien et al. 2004), and in Polar Sweden, the coastal areas are more exposed now and for the future (SMHI 2010). This exposure means increased uncertainty for winter tourism in the coastal areas and raises the question of how best to strategically plan for future tourism in the region. However, in the more stable inland of Polar Sweden, there may be benefits, as in Northern Finland, by gaining a competitive advantage due to relatively poorer conditions in the supply in the European Alps (Saarinen and Tervo 2006). This "gradual shift of tourist destinations towards higher latitudes and altitudes" (Hamilton et al. 2005, p. 263), due in part to a reputation for good conditions (O'Brien et al. 2004), is a positive for winter tourism in Polar Sweden at the regional level albeit with intra-regional differences which should be taken into consideration today. This northward shift in Europe is by no means guaranteed as Alpine communities are also vigorously developing adaptation strategies (Steiger et al. 2011) and other high-altitude winter-tourism markets are emerging, such as Turkey (Demiroglu 2011).

Moen and Fredman (2007) expect substantial monetary losses in the Swedish skiing industry in the long term but under the caveat of static supply and demand in the future. They also list expansion into areas with more reliable snow conditions and development of year-round tourism among future adaptation strategies (Moen and Fredman 2007). The present study assumes a relatively stable demand trend with a change in supply reflecting increased loading of northern winter sport resorts and a concomitant development of year-round tourism in Polar Sweden. It must be noted that strong general development in other sectors of the regional economy could result in decreased vulnerability even as tourism itself grows; thus, an increase in winter tourism is not necessarily indicative of increased vulnerability. The problem for some communities in the polar north is that tourism is seen as the solution rather than a solution among many. The reality, however, is that tourism is far from a panacea for polar areas. This chapter cannot fully explore other avenues of development, but they should be kept in mind throughout.

10.3.3 Tourism as a Complex Adaptive System

> In general, the overall scenario for Northern Europe could be as follows: the summers are getting warmer and drier while more precipitation is expected for the warming winters [. . .]. However, many studies and reports analysing future tourism in the region have been based on generalizations on temperature and precipitation and on snow cover, not thoroughly considering the importance of other possible changes from the viewpoint of tourism. (Tervo-Kankare and Saarinen 2011, p. 223)

The reductionist approach of warmer winters automatically resulting in fewer winter tourists is appealing but offers too simple a picture. Tourism consists of many actors on both the supply and demand side and is more accurately conceptualized as a complex adaptive system. One of the key features of complex adaptive systems is the ability of actors to learn from experience and change behavior so that slowly developing challenges such as climate change are manageable through system adaptations. The complexity comes from the fact that the system has many actors – not all local and not all tourism actors. For example, the issue of adaptation to mitigation policies may be a growing challenge for some regions, particularly those whose remote locations require longer travel due to the correlation between energy use for travel and greenhouse gas emissions (Aall and Höyer 2005; Hamilton et al. 2005). Tourism destinations are both affected by changes in climate but also cause changes in climate, primarily through greenhouse gas emissions resulting from tourist mobility (Gössling 2002). This negative contribution of tourism is germinating increasing government mitigation strategies. Thus, tourism regions, particularly those in peripheral areas, will have an increased necessity for "planned adaptation" (Carter and Kankaanpää 2003, p. 49) to top-down mitigation strategies in the future, a real test of communities' resilience to global changes. In this chapter, changes to local supply are considered as well as changes to local and external demand for winter tourism, but other general developments in the economy are not directly considered here. Instead, the focus is on the local actors who can adapt over time, and this, in the first instance, means the winter-tourism entrepreneurs in Polar Sweden.

10.3.4 Entrepreneurs and Climate Change in Polar Sweden

A concept associated with adaptation and sometimes used in tourism research is resilience (cf. Biggs 2011; Cochrane 2010; Ruiz-Ballesteros 2011; Strickland-Munro et al. 2010). The concept is often taken to refer to "the ability of a system to return to a reference state or remain within a range of desirable states following a perturbation" (ACIA 2004, p. 950) or, in the present study, as the ability of a community to absorb strain placed upon it by exogenous forces. Thus, in the long term, there is no need to assume an ideal state to be preserved, but the important question is whether communities can maintain themselves and evolve to meet changing external circumstances. However, in the medium-term case of winter-tourism development, remaining within a sustainable range translates to having a continuing run of winter-tourism seasons which are economically viable, while the potential perturbation is a significant degradation in climate conditions for winter tourism. Since local resilience is primarily dependent on the autonomous adaptations of winter-tourism entrepreneurs, it is worth looking at how they are thinking about the issue of climate change adaptation (Aall and Höyer 2005; Brouder and Lundmark 2011; Carter and Kankaanpää 2003; Hall 2006; Saarinen and Tervo 2006).

In a recent survey of Polar Sweden's winter-oriented tourism entrepreneurs (n = 63, survey conducted by authors of the present study), the following plans for adaptation to climate change and general previous adaptations were expressed:

- 20% will plan 1 year in advance.
- 60% will plan during the season depending on conditions.
- 40% have had alternative activities in winter.
- 50% have developed summer products.
- 20% have increased capacity.

The fact that most entrepreneurs intend to adapt during the season rather than plan in advance shows that forward planning, if necessary, will have to come from elsewhere, for example, institutions. It also shows the learning through experience inherent in complex adaptive systems – where adaptations will come if and when necessary and not before. A large share of entrepreneurs have had alternative (non-weather dependent) winter activities (40%) or have developed summer products (50%), which further adds to the complexity since these developments may indicate specific responses to changing weather conditions or may indicate legitimate ways to grow the business and increase revenue. In any case, the long-term viability of tourism entrepreneurs in Polar Sweden is good at the moment with many firms making gains in turnover and employee numbers over time (Brouder and Eriksson 2012), and since a large share are engaged in winter tourism, it implies an absolute increase in sensitivity to climate change at the micro-level.

On a more general level, the effects of climate change on demand and supply are expected to relate to changes in tourism demand, for example, the volume of travelers, when people travel during the year, the choice of destination, and the type of holiday experience tourists seek (Aall and Höyer 2005). Thus, a full assessment of vulnerability to climate change, including adaptive capacity, is difficult – without discussing the possible outcome of climate change on tourist behavior, an important part of the system is missing. This is a huge task and beyond the scope of the present study. From other research, it has been concluded, however, that people respond to changes in weather, temperature, and wind conditions (Richardson and Loomis 2004; Denstadil et al. 2011) not to climate changes per se, and so, changing frequencies of inclement weather may be used as a proxy for how people might respond to climate change.

Climate change, tourism, and entrepreneurship overlap in a number of different ways, many of which are complex. Vulnerability is most threatening where resilience is lacking. Vulnerability is a complex phenomenon since not all actors have the same level of vulnerability due to differing exposures or levels of mobility (Brouder and Lundmark 2011). Thus, resilience is a complex term since not all actors in the system will share the same capacity for resilience. However, generating a number of possible scenarios for the future is a good first step which then allows for a richer discussion of the resilience of the various actors.

10.4 Scenarios for Winter Tourism in Polar Sweden in 2040

Based on the above information, it is possible to construct simple, plausible scenarios on what the future may hold for winter tourism in Polar Sweden. Scenarios are, by definition, not conclusive since there are many uncertainties, but they can highlight the likely futures of 2040 and how the development of winter tourism is at a crossroads today. Choosing 30 years hence is a reasonable time frame to consider long-term developments since this is beyond the range of individuals but within a reasonable time frame for institutions and climate models that are "similar through about 2040, but diverge thereafter" (ACIA 2004, p. 994). The scenarios focus on winter-tourism changes only.

Following van Notten et al.'s (2003) scenario typology, the scenarios presented here are exploratory with a medium-term (three decades forward) time scale and a regional (Polar Sweden) spatial scale. The scenario data are drawn from the following sources: climate – ACIA (2004), Lind and Kjellström (2008), and SMHI (2010); demand – Hall and Saarinen (2010), Hamilton et al. (2005), and Lundmark (2010); and supply – Brouder and Lundmark (2011), Moen and Fredman (2007), and Saarinen and Tervo (2010). There was, however, some crossover in the source texts between supply and demand, and so, the research process involved much cross-checking of texts and qualitative interpretation by the authors. The climate data were used to construct scenarios for the winter-tourism sector that are based on (1) the most likely (scenario 1), (2) the least impacting but possible (scenario 2), and (3) the most impacting but possible (scenario 3). Thus, the final scenarios offer three possible snapshots of the future of winter tourism in Polar Sweden with a restricted level of deviation due to the reliable climate models available for 30 years hence (Table 10.1). Since the scenario construction is an iterative process, the experience of the authors in the region is also evident in the final output; however, a final check was made to ensure that nothing in the final scenarios contradicts the source texts.

10.5 Adaptation to Climate Change Within the Arctic Tourism Innovation System

An innovation systems approach is a natural lens through which to view the development of winter tourism in Polar Sweden since innovation is multi-actor driven and so is adaptation. Tourism innovation systems include the social, economic, institutional, cultural, and regulatory environments. The Arctic Tourism Innovation System (ATIS) framework allows an exploration of winter-tourism and climate change adaptations within the particular milieu of the Arctic by utilizing a unified sectoral and regional approach (Lundmark and Müller 2012). Under the ATIS framework, climate change may be conceptualized as a challenge to communities, one which requires innovative responses by various actors (entrepreneurs and

Table 10.1 Basic scenarios for winter tourism and climate change in Polar Sweden in 2040

Scenario	Climate	Demand	Supply
1. Consolidation	Winter and spring temperatures increase by 2°C on average Increase in variability noticeable in the coastal areas with later initial snowfalls shortening the potential season	Increasing slightly, mostly interregional tourists, as climate change mitigation policy increases air travel costs Fewer day-trips for skiing in coastal hills	Continued consolidation in the inland resorts Similar supply on coast but growth among mobile operators only
2. Status quo	Winters not more than a fraction of a degree warmer Lots of snow with no noticeable decrease in snow cover even in the coastal areas Some winters with early, heavy snowfalls	Increase in demand, mostly from Nordic countries but also from other regions of Europe Slight increase in day-trips to coastal hills especially in winters with earlier snowfall	Some new supply in the inland but growth tending to cluster around larger businesses Similar supply on coast but with modest increase in both venue-based and mobile operators
3. Encroachment	Coastal winter season shortens by more than one month on average Inland sees increased precipitation with more snow but also more extreme spells	Increasing numbers of Scandinavian tourists in the Swedish mountain range European market diverting to polar areas due to problems in the Alps but travel costs increase significantly	Fewer viable firms in the coastland and those remaining are mostly mobile operators Continuing supply in inland, mostly resort-based with other activities for days with extreme conditions

Note: Scenarios based on ACIA (2004), Brouder and Lundmark (2011), Hall and Saarinen (2010), Hamilton et al. (2005), Lind and Kjellström (2008), Lundmark (2010), Moen and Fredman (2007), Saarinen and Tervo (2010), and SMHI (2010)

institutions) within the local system. This approach allows the challenge of climate change to be seen as not just a tourism business concern but as a coming challenge for polar communities. It also allows the complex and interconnected actors in the tourism system to be included in the search for innovative solutions to the challenges posed by climate change. Adaptation to climate change is thus closely linked to the ATIS as the preconditions for innovation largely coincide with adaptive capacity on different geographical scales.

Private business endeavors are inherently linked to public institutions and strong support policy apparatus for planned adaptation involving international to local levels (Storper and Scott 1995; Hall 2011). Who is positioned to lead and make the necessary planning decisions today? Hjalager (2002) states that it is not measures immediately directed toward the industry that are crucial for innovation. Instead, policy measures should take into account the driving forces of, for example, the public sector for increasing innovation and consequent adaptation (Hjalager 2002) and distribute resources accordingly in order to optimize the innovative capacity of the system.

In the case of climate change, it is thus the regional institutions which must lead since research in several regions has shown that the entrepreneurs in general are unwilling or unable to engage in long-term planning for climate change. The importance of regional institutions is invoked by the "tragedy of the commons" concept (Hardin 1968); that is, left to their own devices, entrepreneurs may succeed in developing in the medium term without realizing they are increasing their sensitivity, and if this occurs in an increasingly exposed area, then the vulnerability equation may change palpably, thus making any future adaptation more difficult than it would otherwise need to be. It is also important to note that there are secondary effects within the tourism economy as support services grow around winter-tourism development and may become increasingly reliant upon the winter-tourism economy over time. Therefore, the role of institutions in supporting sustainable development for, in the medium term, entrepreneurs and, ultimately, communities cannot be overlooked.

Polar Sweden faces an uncertain future but, in any case, a contraction of the season, and increased variability during the season is likely. The general tendency for further urbanization might affect tourism development as well. Outdoor recreation in urban proximate areas could develop into nature-based tourism for an increasingly urban population demanding sportified tourism products in safe "natural" environments and the development of facilities that replicate ideal conditions and facilitate access to outdoor recreation might reduce the effect of climate change on tourism demand for some geographical locations. Lundmark and Müller (2010) conclude that for the polar north, including Polar Sweden, increasing tourism can only be achieved through inbound tourism. This adds a further nuance to the question of sensitivity – tourists coming from outside the nearest regions are subject to a wider set of conditions and opportunities, due in part to the hierarchical flexibility within the tourism system (Lundmark 2010): the tourist is most flexible, and thus, if the destination does not meet their preferences, the tourist will go somewhere else. The travel organizers are less flexible although they are able to change their supply in a season or two. Small-scale, locally owned, nature-based tourism businesses are most vulnerable to climate change (Saarinen and Tervo 2006). This is a major question mark for the future under *scenario 3 – encroachment* as the increasing cost of long-distance travel will impinge on the demand side and will test the price elasticity of demand for winter tourism.

A recent survey of Polar Sweden's winter-tourism businesses ($n = 63$, survey conducted by authors of the present study) found that 71% of businesses offer no snow guarantee. The region is snow secure under both *scenario 1 – consolidation* and *scenario 2 – status quo,* and the entrepreneurs in the survey stated that within 10 years, winter tourism will still be viable. The lack of guarantees offered is due to a combination of micro-businesses and largely domestic markets which mean this type of product augmentation is either not considered or deemed unnecessary. Taking an interregional perspective, one can immediately see the potential competitive advantage of a relatively snow-secure region promoting itself as such, and even from an intra-regional perspective, going forward, offering a snow guarantee may become more important as the coastal population may not see suitable winter-tourism conditions around them and may then not consider a winter sport holiday, even if the conditions in the interior supply area remain stable. As much of Polar Sweden remains stable, now and for the near future, it is important to highlight this fact in order to promote winter tourism. However, guaranteeing snow is not the same as guaranteeing suitable activity-specific conditions, which have quite a variance within winter tourism (Tervo 2008), and snow guarantees in the early winter season can prove expensive if one poor-snow winter results in large trip refund payments.

Danish economist Ester Boserup stated that in times of distress, humans innovate to adapt to changes (Boserup 1965), and although she was referring to food shortages due to agricultural limits, her idea could also apply to scarce economic opportunities in other sectors. Ultimately, the adaptation of entrepreneurs will occur, but not before there is a palpable stress on the local tourism economy (Biggs 2011), so *scenario 3 – encroachment* is the only one which would result in major adaptations with *scenario 1 – consolidation* and *scenario 2 – status quo,* causing changes which are harder to discern from non-climate-related motivations. Regional institutions have the opportunity to take a longer-term view and can adjust their planning so that vulnerable regions are not put on a path that will bring greater stress in the future. This does not mean an abandonment of winter tourism in certain areas within Polar Sweden but rather a more mature approach to community planning which bears in mind the longer-term scenarios while addressing the shorter-term needs of the communities.

10.6 Conclusion

According to triangulation of different climate models, it is quite clear how the climate will be, within a given range along the dimensions of both snow cover and snow conditions, but it is more difficult to know what society will be like in 2040 – what changes in other aspects such as cultural and regulatory systems will have occurred? One thing is clear, however, change is inevitable. Only through regional adaptation planning which considers the local geographical context will Polar Sweden be equipped to face the uncertain future. It is the human/environment interaction which decides the exposure and sensitivity of regions, but it is policy and planning at the regional level which determines adaptive capacity. This chapter calls for a greater connection between innovation systems theory and long-term adaptation to climate change in polar tourism research.

Future studies should detail the linkages between adaptation to climate change and the Arctic Tourism Innovation System (ATIS) through theoretical development and empirical assessment of current processes of adaptation and innovation in general and, in particular, related to tourism and climate change in the polar north. The special characteristics of high-latitude regions need to be considered in future innovation systems research. Tourism-related studies of the polar north can no longer ignore the intra-regional geographical relationships since higher-resolution climate modeling and particular community conditions demand they be considered.

Acknowledgements This chapter has been part-funded by the research program Outdoor Recreation in Change (Friluftsliv i förändring), which is financed by the Swedish Environmental Protection Agency (Naturvårdsverket), and by the Swedish Foundation for Strategic Environmental Research (MISTRA). The authors hereby acknowledge the economic support of both agencies.

References

Aall C, Höyer GK (2005) Tourism and climate change adaption: the Norwegian case. In: Hall CM, Higham J (eds) Tourism, recreation and climate change. Channel View Publications, Clevedon
ACIA (2004) Impacts of a warming arctic: arctic climate impact assessment. Cambridge University Press, Cambridge
Becken S, Hay J (2007) Tourism and climate change: risks and opportunities. Channel View Publications, Clevedon
Biggs D (2011) Understanding resilience in a vulnerable industry: the case of reef tourism in Australia. Ecol Soc 16(1):30
Boserup E (1965) The conditions of agricultural growth: the economics of agrarian change under population pressure. George Allen & Unwin publishing, London
Brouder P, Eriksson RH (2012) Staying power: what influences micro-firm survival in tourism? Tour Geogr 15(1)
Brouder P, Lundmark L (2011) Climate change in Northern Sweden: intra-regional perceptions of vulnerability among winter-oriented tourism businesses. J Sustain Tour 19(8):919–933
Carter T, Kankaanpää S (2003) A preliminary examination of adaptation to climate change in Finland. Finnish Environment Institute, Helsinki
Cochrane J (2010) The sphere of tourism resilience. Tour Recreat Res 35:173–186
Demiroglu C (2011) Impact of climate change on winter tourism: a case of Turkish Ski resorts. Kulturgeografiska institutionen, Umeå
Denstadil JM, Jacobsen J, Lohmann M (2011) Tourist perceptions of summer weather in Scandinavia. Ann Tour Res 38(3):920–940
Gössling S (2002) Global environmental consequences of tourism. Glob Environ Chang 12:283–302
Hardin G (1968) The tragedy of the commons. Science 162(3859):1243–1248
Hall CM (2006) New Zealand tourism entrepreneur attitudes and behaviours with respect to climate change adaptation and mitigation. Int J Innov Sustain Dev 2:229–237
Hall CM (2011) Policy learning and policy failure in sustainable tourism governance: from first- and second-order to third-order change? J Sustain Tour 19(4–5):649–671
Hall CM, Saarinen J (2010) Last chance to see? Future issues for polar tourism and change. In: Hall CM, Saarinen J (eds) Tourism and change in polar regions: climate, environment and experience. Routledge, London, pp 301–310
Hamilton JM, Maddison DJ, Tol RSJ (2005) Climate change and international tourism: a simulation study. Glob Environ Chang 15(3):253–266
Hjalager AM (2002) Repairing innovation defectiveness in tourism. Tour Manage 23:465–474
Lind P, Kjellström E (2008) Temperature and precipitation changes in Sweden: a wide range of model-based projections for the 21st century. SMHI-RMK 113. Swedish Meteorological and Hydrological Institute, NorrkoÌĹping
Lundmark L (2005) Economic restructuring into tourism in the Swedish mountain range. Scand J Hosp Tour 5(1):23–45

Lundmark L (2010) A holiday on ice on hold? Nature-based tourism and climate change in the Nordic north. In: Hall CM, Saarinen J (eds) Tourism and change in polar regions: climate, environment and experience. Routledge, London, pp 135–146
Lundmark L, Müller DK (2010) The supply of nature-based tourism activities in Sweden. Tourism 58(4):379–393
Lundmark L, Müller DK (2012) Arctic tourism innovation system (ATIS). Abstract for the 3rd International Polar Tourism Research Network conference, Nain, Canada
Moen J, Fredman P (2007) Effects of climate change on alpine skiing in Sweden. J Sustain Tour 15(4):418–437
Müller DK (2011) Tourism development in Europe's "last wilderness": an assessment of nature-based tourism in Swedish Lapland. In: Grenier AA, Müller DK (eds) Polar tourism: a tool for regional development. Presses de l'Université du Québec, Québec, pp 129–153
Müller DK, Jansson B (eds) (2007) Tourism in peripheries: perspectives from the far north and south. CABI publishing, Wallingford
Noakes JL, Johnston ME (2009) Constraints and opportunities in the development of diamond tourism in Yellowknife, Northwest Territories. In: Hall CM, Saarinen J (eds) Tourism and change in polar regions: climate, environment and experience. Routledge, London, pp 165–179
O'Brien K, Synga L, Haugen J (2004) Vulnerable or resilient? A multiscale assessment of climate change impacts and vulnerabilities in Norway. Clim Chang 64:193–225
Richardson R, Loomis J (2004) Adaptive recreation planning and climate change. A contingent visitation approach. Ecol Econ 50:83–99
Ruiz-Ballesteros E (2011) Social-ecological resilience and community-based tourism: an approach from Agua Blanca, Ecuador. Tour Manage 32(3):655–666
Saarinen J, Tervo K (2006) Perceptions and adaptation strategies of the tourism industry to climate change: the case of Finnish nature-based tourism entrepreneurs. Int J Innov Sustain Dev 1(3):214–228
Saarinen J, Tervo K (2010) Sustainability and emerging awareness of a changing climate: the tourism industry's knowledge and perceptions of the future of nature-based tourism in Finland. In: Hall CM, Saarinen J (eds) Tourism and change in polar regions: climate, environment and experience. Routledge, London, pp 147–164
Schneider S, Sarakhun J et al (2001) Overview of impacts, adaptation, and vulnerability to climate change. In: McCarthy JJ, Canzizni OF et al (eds) Climate change 2001: impacts, adaptation, and vulnerability. Cambridge University Press, Cambridge, pp 75–103
SMHI (2010) Klimat, observationer, och framtidsscenarier – medelvärden för länet (Norrbottens län & Västerbottens län), compiled 7 December 2010
Smith K (1990) Tourism and climate change. Land Use Policy 7(2):176–180
Steiger R, Dawson J, Stötter J (2011) Last chance tourism in alpine regions: last chance to ski? In: Lemelin RH, Dawson J, Stewart E (eds) Last-chance tourism: adapting tourism opportunities in a changing world. Routledge, London, pp 42–54
Storper M, Scott A (1995) The wealth of regions. Market forces and policy imperatives in local and global context. Futures 27(5):505–526
Strickland-Munro JK, Allison HE, Moore SA (2010) Using resilience concepts to investigate the impacts of protected area tourism on communities. Ann Tour Res 37:499–519
Tervo K (2008) The operational and regional vulnerability of winter tourism to climate variability and change: the case of the Finnish nature-based tourism entrepreneurs. Scand J Hosp Tour 8(4):317–332
Tervo-Kankare K, Saarinen J (2011) Sub-arctic environmental dimension: climate change and adaptation strategies of the tourism industry in Northern Europe. In: Maher PT, Stewart EJ, Lück M (eds) Polar tourism: human, environmental and governance dimensions. Cognizant Communications Corporation, New York, pp 221–235
Van Notten PWF, Rotmans J, Van Asselt MBA, Rothman DS (2003) An updated scenario typology. Futures 35:423–443

Chapter 11
Conflicting Discourses in Tourism Development

A Case of Kilpisjärvi Village, Finnish Lapland

Tuukka Mäkitie and Anna-Liisa Ylisirniö

Abstract Kilpisjärvi is the northernmost village in western Finnish Lapland, located in the middle of the highest mountains of Finland. The region has been reindeer herding area of Sámi people for centuries but has also attracted skiers and hikers since 1930s. During the last few decades, tourism has increased strongly in the area, comprising now about 1,400 beds. We studied the discourses of tourism development and their underlying values by interviewing permanent dwellers and Norwegian cottage owners. Six different discourses were distinguished: (1) nature and scenery, (2) economy, (3) transformation as a place, (4) reindeer herding, (5) planning and administration, and (6) possibilities to influence decision-making. The first discourse stressed that tourism development and building should adapt to fragile nature, while economic discourse stressed tourism as the most important means of livelihood in the village. Contradictory values and identities of the place were expressed among the respondents, some seeing the beauty and close-knit community to be lost and others seeing the transformation positive. Reindeer herding contradicted tourism strongest, having both areal and cultural conflict. Governance was seen as an important tool in steering development, but many respondents considered that it was taking into account only economic aspects. Many villagers saw their influence in decision-making weak. We conclude that it is important to include the social implications and the diversity of local discourses in

T. Mäkitie
Department of Geography, University of Helsinki, Helsinki, Finland

Department of Development and Planning, Aalborg University, Postboks 159, 9100 Aalborg, Denmark
e-mail: tuukka.makitie@gmail.com

A.-L. Ylisirniö, Ph.D. (✉)
Arctic Centre, University of Lapland, POB 122, 96101 Rovaniemi, Finland
e-mail: anna-liisa.ylisirnio@ulapland.fi

D.K. Müller et al. (eds.), *New Issues in Polar Tourism: Communities, Environments, Politics*, DOI 10.1007/978-94-007-5884-1_11,

tourism development and land use planning. Transparent and genuine participatory planning would promote sustainability and allow decisions and development which would satisfy larger parts of local population than presently.

Keywords Tourism development • Conflicting discourses • Local community • Land use planning • Participatory planning

11.1 Introduction

As nature-based tourism is gaining in popularity and more remote and environmentally fragile destinations have become the focus of large-scale tourism, also the interest of tourism research has turned to the peripheral areas. Even though the local communities around peripheral tourism destinations are usually small, they should not be overlooked when tourism is studied, planned, and developed. To avoid or ease conflicts between different stakeholders, we need information on the social formations of nature-based tourism development at the local level.

Initially, the main interest in the field of nature-based tourism research concentrated on the impacts of visitors on nature, while sociocultural and economic impacts received less attention. Later, the research community has responded to the need to understand the social dimensions of nature-based tourism at the local level by studying the controversies between tourism development and local communities (e.g., McDonald 2009; Puhakka et al. 2009; Choi and Murray 2010; Dredge 2010; Haukeland 2011). Dredge (2010), for instance, noted that powerful interest groups and narrow economic interpretations of place can dominate tourism development discourses. Also Hall (2008: 65) has stated that free market "is not an adequate mechanism by itself to protect the interests of all parties and stakeholders in tourism development process." Therefore, as Castellani and Sala (2010) have pointed out, it is necessary to widen the involvement of local residents in the planning processes from mere consultancy to active participation. However, it should be noted that this is not a simple process. Gunn and Var (2002: 339–340) and Cooper et al. (2005: 317) reminded us about the complexity of the perceptional sphere in any tourist destination: All people experience, understand, and value the tourist destination from their own point of view. During the past few years, an increasing number of studies have also concentrated on the management tools and policies to solve current tourism development issues (e.g., McCool 2009; Plummer and Fennell 2009; Wray 2009; Logar 2010). For example, McCool (2009) addressed the problem of powerful interest groups and narrow economic interpretations of tourist destination by arguing for the need to construct sustainable tourism with no single public interest. Statements have been made (e.g., Jokinen and Sippola 2007; Saarinen 2007: 49, Mettiäinen et al. 2009) that local stakeholders should be more thoroughly integrated to the planning process in order to achieve a more sustainable situation in tourism development in Finnish Lapland. In Finland, the Land Use and Building Act (1999) states that plans "must be prepared in interaction with persons and bodies on whose

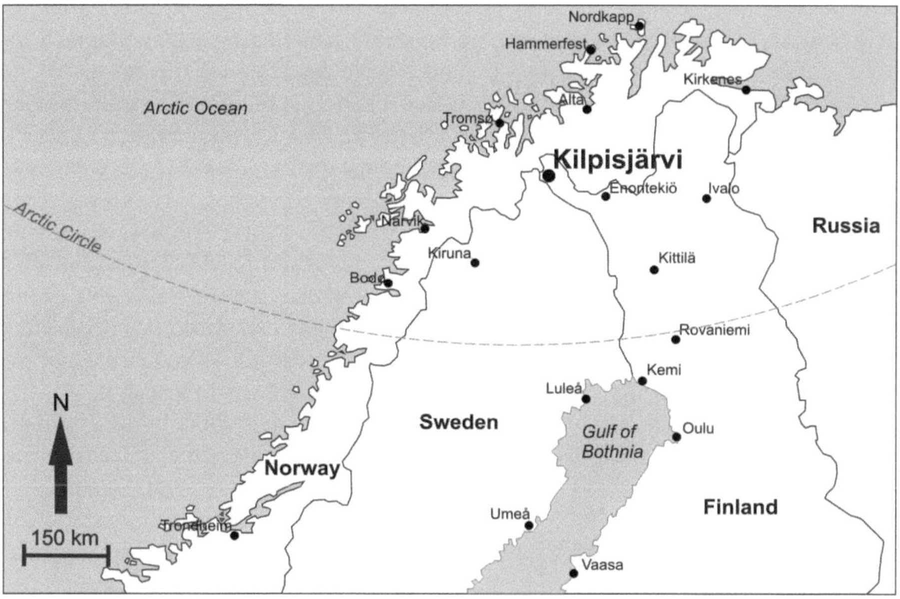

Fig. 11.1 Location of Kilpisjärvi village

circumstances or benefits the planning may have substantial influence." A case study of three tourist villages in Finnish Lapland reveals that there are different experiences how the community participation process has succeeded. The results showed that when taken into account, the local knowledge of permanent residents could improve the planning process (Tuulentie and Mettiäinen 2007). McDonald (2009) has stated that appropriate planning and sustainable tourism development can be achieved through the understanding of stakeholders and components of the study area. Our study is designed to contribute to this task by creating a case study about local level discourses and thus creating means to understand this and other similar cases.

The purpose of this chapter is to study the perceptions, statements, and conflicting discourses of the physical tourism development in the village of Kilpisjärvi, Finnish Lapland. This remote village with approximately 100 permanent residents is located in northwestern Finnish Lapland (Fig. 11.1), bordering both northern Norway and Sweden. The Sámi people have used the area as reindeer pastures and summer settlement for centuries, and they still use the area even though tourism development is partly occupying the traditional herding areas. Permanent settlement in the area was established only about 80 years ago. Nature-based tourism in Kilpisjärvi is almost as old as the village itself. Initially, tourism was small scale with one hotel and a few basic accommodations until the turn of the millennium, when the area experienced an explosion of tourism development, especially in the form of upscale second homes, which were built and sold mostly to Norwegians nearby. This change also included the transformation of the typical tourist type of Kilpisjärvi: Middle-aged Finnish backpackers and cross-country skiers have been replaced to a great extent by younger Norwegian snowmobile drivers. Both groups are still using the same areas, which creates conflicts among these user groups (see Arnegger et al. 2010 for comparison). The accommodation capacity of Kilpisjärvi was multiplied in just a decade, being around 1,400 beds in 2009, and new accommodations are currently planned. This development has raised concerns both locally and nationally, as the scenery of Kilpisjärvi with the highest mountains in Finland, as well as rare flora and fauna, is considered nationally unique. The Saana Mountain, which is the main tourist attraction near the village, is one of the most recognizable mountain silhouettes in Finland. The development of Kilpisjärvi has given rise to many controversial statements, both in favor and against the current development, thus making it an interesting case study for tourism research.

Our approach is based on the assumption that the local community of a tourism destination attaches values and meanings to their surroundings, thus creating meaningful places (Tuan 1974, 1977) and landscapes (Cosgrove 1984; Daniels and Cosgrove 1988; Matless 1998). We assume that these perceptions of space differ from one individual to another; therefore, attitudes and opinions vary within a local community, for instance, on land use and development issues. Through the interviews of local residents, we aim at investigating how rapid tourism development is socially constructed by the local community, discovering the main discourses and most controversial issues, and suggesting ways of reducing conflicts in tourism destinations in a similar phase of development.

11.2 Methodology

We conducted 20 qualitative research interviews in Kilpisjärvi during the summer 2009. Seventeen of the interviewees were Finnish residents of Kilpisjärvi or nearby areas, and three were Norwegian cottage owners. The process of selecting the interviewees started by making an overview on the social and demographic structure of the permanent and semipermanent residents of Kilpisjärvi and then by selecting interviewees to represent as evenly as possible the different gender, age, and occupational and stakeholder groups of the village. Two of the interviewees were Sámi reindeer herders. Most of the preselected interviewees agreed to contribute to the study. Each interview was based on a standard list of questions, but interviewees were also allowed to speak freely on any theme they chose regarding Kilpisjärvi and tourism development. All interviews were recorded and transcribed.

The methodological framework of the study is discourse analysis, which positions the statements of the interviewees in their larger discursive contexts, defining what is actually said in the speech act (Calhoun 2011). The approach is based on the view that the analysis reveals what kind of issues have social importance because written or spoken discourses do not represent only individual opinions but bear more extensive cultural meanings (Berger and Luckmann 1966). We used discourse analysis as a framework to find answers to two main research questions: What are the main elements of discussion on tourism development in Kilpisjärvi, and how are they addressed? From the various perceptions and points of view raised in the interviews, we analyzed those main elements which cut through different interviews and could thus be interpreted as independent discourses. Through understanding the variant discourses of the topic at hand, we are enabled to look upon the complexity of tourism development at a local level.

11.3 Results

Six different discourses on tourism development in Kilpisjärvi could be distinguished from the interviews (the order of discourses does not suggest any hierarchy or differences in level of importance): (1) nature and scenery, (2) economy, (3) transformation as a place, (4) reindeer herding, (5) planning and administration, and (6) possibilities to affect decision-making. In the following, these discourses are discussed in more detail.

11.3.1 Nature and Scenery

The nature and scenery of Kilpisjärvi were highly respected among the interviewees and were mentioned often in the aesthetic sense. The residents deeply admired their surroundings and the scenery of Kilpisjärvi. For instance, it was common to call Kilpisjärvi one of the national landscapes of Finland, even though, in the official sense, this is not the case (the Finnish Ministry of Environment has defined 27 places in Finland to be the national landscapes of the state; Kilpisjärvi is not one of these). It was widely agreed that tourism development should adapt to the scenery of Kilpisjärvi. For instance, the interviewees stated that the architecture of new buildings should fit into the natural landscape and the construction sites should be left as neat as possible. In other words, visual factors seemed to play a key role in the acceptability of housing projects. Many respondents did not consider the previous housing projects successful since the new buildings were not architecturally cohesive, resulting in a mosaic appearance of the village. In addition, some of the contemporary buildings that do not follow the traditional local way of building were described as "grotesque" and "hideous" by local residents.

Some respondents expressed concern about the carrying capacity of the environment. They demanded more discreet and careful tourism development in order to avoid long-lasting damages to the nature under the pressure of the Arctic climate and short growing season. In other statements, it was noted that Kilpisjärvi should be treated delicately, and the uniqueness of Kilpisjärvi should be perceived. In some statements, it was also claimed that the municipal administration (the land use authority according to Finnish law) did not possess enough knowledge and know-how to deal with land use development issues in environmentally fragile areas such as Kilpisjärvi. Some second homeowners of Norwegian background also raised concerns regarding these developments.

11.3.2 Economy

Tourism-based services provide a livelihood for many residents in Kilpisjärvi, and they are also an important source of income to the municipality of Enontekiö. Therefore, it was obvious that economic themes would come up when discussing tourism development. As a tourist attraction, nature is perceived as a valuable resource and potential economic generator. Statements used in the economic discourse emphasized, for instance, that there was no unemployment in the village. Terms such as "the entrepreneur village" were also used. It was widely accepted among the respondents that tourism is a part of Kilpisjärvi, and there were no major controversies against the phenomenon as such. There was also a willingness to accept damages to scenery (even though they were deprecated, as seen in the previous discourse) as a necessary sacrifice for much-needed economic gains in the village. Tourism entrepreneurs had mainly positive attitudes toward tourism development, as the increase in the volume of tourism boosts their own businesses directly.

However, a few complaints were heard among the respondents. Some of them argued that only a few people benefited from tourism development, while housing projects "spoiled" the scenery for everybody. The landscape of Kilpisjärvi was considered as common property, and, while only few were making money out of it, all were condemned to endure the negative consequences of this development. The municipality of Enontekiö was also accused of acting too hastily in building projects and being interested only in the economic spin-offs from tourism. In fact, Enontekiö, which is one of the poorest municipalities in Finland, has been very active in the tourism development of Kilpisjärvi. Nevertheless, some interviewees claimed that this pursuit was not soundly planned and was driven by the need to gain income in the short term.

Economic progress appeared as an independent value among some of the respondents. Especially the entrepreneurs of the village expressed the need for economic growth in order to remain competitive compared to other tourist destinations in Lapland. On the other hand, there were concerns that the rapid development would prove to be short sighted and unsustainable in the long term. According to the perceptions of these respondents, the "business way of thinking" did not fit Kilpisjärvi, and they claimed that the true spirit of Kilpisjärvi should not be bound to the laws of economy and profit.

11.3.3 Value and Transformation as a Place

One distinguishable discourse in tourism developments pertained to the changes which had occurred in Kilpisjärvi. Some respondents looked back to the village Kilpisjärvi as what it used to be: a remote and quiet place in the midst of a northern wilderness. They perceived the village differently nowadays. Noisy snowmobiles and increased housing have transformed Kilpisjärvi into a tourist resort, not a peaceful village in Lapland. A few claims were also made that the development had affected the spirit of the village: The once closed-knit and balanced community was no longer the same, and tourism developments and land use issues had replaced the more peaceful era of the twentieth century. Even though the development was generally regarded as a normal phenomenon, as noted in the previous discourse, there were concerns that the change in the spirit of Kilpisjärvi had gone too far. The respondents feared that the attractiveness of the area would decrease if Kilpisjärvi was developed too much further. Some of the interviewees claimed that critical errors had already been made and that Kilpisjärvi had lost its reputation as a resort where one can experience the peace of nature.

Some respondents held an opposite view noting that change is always inevitable and that one should not be bound to the past. Therefore, the progress and transformation of a place is natural, and the villagers should concentrate on what the best direction for the change would be. Some entrepreneurs also stated that Kilpisjärvi draws a lot of local and nationwide attention due to its natural uniqueness on a national scale. From this point of view, many of the accusations of spoiling the natural beauty of Kilpisjärvi are therefore undeserved, as the same actions to develop tourism would be considered business as usual in other locations in Lapland. This perception of unfairness is part of the larger controversy between local and expert knowledge. Due to its specific nature, Kilpisjärvi has drawn academics through decades, and some local residents felt that they could not influence their own surroundings, as "experts from the south" had more knowledge and power over the area than the locals. This controversy between local knowledge and expert knowledge is also bound to the typical locals vs. outsiders dichotomy that can be found in different degrees in Kilpisjärvi and in other parts of Lapland.

11.3.4 Reindeer Herding

Reindeer herding is the oldest form of land use in the Kilpisjärvi area; large-scale reindeer herding started in western Lapland by Sámi people in the sixteenth century (Itkonen 1948). Before the border was closed in 1852, Sámi families traveled between the Norwegian coastline and northwestern Finnish Lapland, where the winter pastures located (Korpijaakko-Labba 2000). Kilpisjärvi region has traditionally been a summer pasture and a place for herders to meet other Sámi families from nearby areas. Unlike in Sweden and Norway, the right to herd reindeer in Finland is not restricted to Sámi, but in Kilpisjärvi region it has been practiced by the same Sámi families for centuries (Heikkinen et al. 2005).

The discourse of reindeer herding differed from other discourses: Reindeer herding is not only a livelihood but it is also an integral part of the Sámi culture and a way of life. The interviewed Sámi reindeer herders saw the area from a reindeer herder's perspective, concentrating on where reindeer find food and what kind of landscapes the animals favor. Unfortunately, it seems that both reindeers and tourism tend to favor the same patches of land. The housing in Kilpisjärvi, even though relatively small in volume, is concentrated on the lushest and most fertile areas around Lake Kilpisjärvi which are also the best grazing areas for reindeer. The situation is complicated because reindeer herding is not allowed in Malla Strict Nature Reserve, which is located in the northwestern side of the village. The new tourist accommodations planned in the northern part of the village would disconnect the current grazing area around Saana Mountain (Fig. 11.2), still decreasing traditional grazing areas. Increasing tourism development also means more distractions to herding, as the animals are shy. This means more work for the herders to keep the herd together. Therefore, the reindeer herders saw tourism development as a conflicting development for their culture. The respondents were concerned that the weight of reindeer herding was decreasing in the local decision-making and that herding as a means of livelihood was pushed aside due to its smaller economic value compared with tourism.

11.3.5 Planning and Administration

According to the Land Use and Building Act (1999), "Land use in municipalities is organized and steered by local master plans and local detailed plans. The local master plan indicates the general principles of land use in the municipality." Thus, the municipality is in a key position in local land use issues, as it has the right to plan and zone its territory according to the guidelines of provincial land use plans.

Land use planning has been a hot topic in Kilpisjärvi for a long time. As seen in previous discourses, Kilpisjärvi is considered a precious location with its values worth protecting. However, insights on how much and what kind of planning is enough vary among the respondents, and this discourse was actually fueled with dissatisfaction on the actions of the local administration. Claims were made that the municipality was not capable of dealing with the complex land use issues in Kilpisjärvi. According to these opinions, the landscape issues pertaining to Kilpisjärvi were not taken into account in the land use planning, and the municipality was in fact not strong enough to implement the plans. It was stated that the municipality was actually too eager in its support of development projects in or near the village and that it did not bother to oblige the landowners to follow the land use plans of the area. Some respondents wondered especially how the municipality of Enontekiö had not imposed any penalties for obvious violations

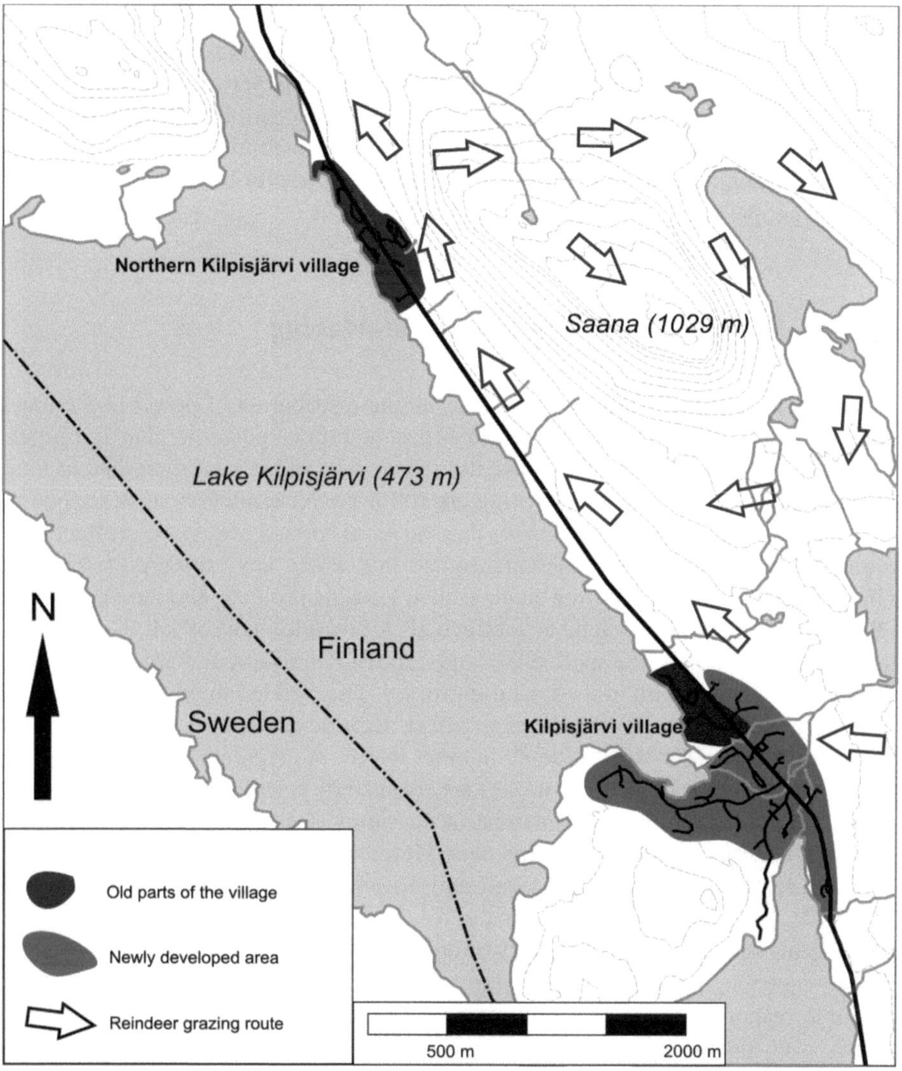

Fig. 11.2 Tourism development areas of Kilpisjärvi and reindeer grazing routes around Saana Mountain during summer months. The planned new tourist resort in the northern part of the village threatens to disrupt the traditional grazing route of reindeers on the northwestern slope of Saana

of land use regulations. The most common example was the clearance of a new camping site by a local entrepreneur. The site was established on an ecologically and visually important area at the root of the Saana Mountain. The mountain is ecologically special due to the calcareous soil with red-listed plant species and rare invertebrates and birds, and the cleared site is visible from far away on the slope of the mountain. The clearance was carried out without a formal procedure and permission, and it also drew some national attention in the media. Nevertheless, no consequences were issued, and the situation continued as business as usual. This has been one of cases that have increased the doubts on the expertise and capability of the municipal decision-makers to deal with a complex location such as Kilpisjärvi. There were even cries for the intervention of national institutions on the municipal decision-making, but first and foremost, there were hopes for better participation of the local people.

11.3.6 Possibilities to Affect Decision-Making

A consistent emerging discourse emphasized the distribution of power in the local land use issues. As noted, the municipality has the official power to plan and supervise land use issues. Landowners have the self-evident right to use their land as long as the regulations of land use planning are followed. Nevertheless, as described in previous discourses, there were claims that the plans had not always been effectively implemented. Some of the residents argued that a few key entrepreneurs in the village have had completely free hands in their housing projects, and the supervision of construction has been vague or ineffectual. A few interviewees felt that some of these housing projects had spoiled the landscape of Kilpisjärvi and argued that their voice has not been heard in the land use process. This had led to wishes for broader participation and better possibilities to affect decision-making. A common notion was "they hear us, but won't listen"; in other words, the regulations for hearing the stakeholders were followed, but no serious consideration was given to their opinion. As previously stated, the main interest of the municipality, the land use authority, had been to support tourism development. However, some of the respondents were more interested in, for instance, developing Kilpisjärvi as a pleasant village to live in and not, for example, in increasing the volume of tourism. Therefore, some of the respondents accused the municipality of only following the economic discourse and not paying enough attention to the alternative visions of Kilpisjärvi.

Some respondents felt that only entrepreneurs' voices were heard in land use issues, even though many residents in Kilpisjärvi were not economically bound to tourism. Therefore, the economic discourse which was discussed earlier is, to some extent, in conflict with the governance discourse. The respondents demanding more opportunities to participate felt also that other than economic values should be respected in decision-making. One of the second homeowners also wanted more influence for house owners, arguing that their group has invested a lot in the area, and should therefore have a say in the planning process. On the other hand, some entrepreneurs stated that the participation process was already making the development too difficult, as the "complainers" were allowed to slow down land use planning processes. This had taken place in the process of deploying a town plan to the northern part of the village, which would allow a dramatic increase in the volume of tourism-related housing near Saana Mountain. This process was delayed by a complaint submitted by a local environmental nongovernmental organization.

Among the arguments against the tourist resort in the northern part of the village were its expected negative impacts on the possibilities to practice reindeer herding in the area and the destruction of some rare plants and animals.

11.4 Discussion and Conclusions

We discovered six distinguishable discourses on the tourism development of Kilpisjärvi: nature and scenery, economy, value and transformation as a place, reindeer herding, planning and administration, and possibilities to affect decision-making. Some of these discourses are overlapping, but some represent undeniably conflicting views. As stated earlier in Hall's (2008) and Dredge's (2010) discourses, tourism destinations are often characterized by strong financial aspirations which cannot encompass all the interests of all stakeholders. This was also the case in Kilpisjärvi. One of the problematic issues was the challenge to fit together the economic development and the nature and scenery of Kilpisjärvi. The area is known for its natural beauty, and for some respondents, the increasing development was destroying the aesthetic values of the area. In other words, nature was a way to see Kilpisjärvi (see also Tuulentie 2004). Nevertheless, tourism is an important source of income in the area, and development was seen as necessary in order to remain competitive against other tourist destinations. This attitude brought about a willingness to tolerate visual transformations in order to gain economic benefits (cf. Puhakka et al. 2009). Therefore, the support for tourism development is an integral part of the discourse. However, it should be noted that economic factors are only one piece in the puzzle. Many respondents stated that the economic discourse had played a dominant role in decision-making in the past, and following a single discourse was not a sustainable way to develop the tourism of Kilpisjärvi. Even though nature and sceneries, together with economic and labor capacities of the village, could be seen together as development resources of Kilpisjärvi, at the moment they were seen rather as diverse lines of thought, being in conflict with each other. For both ecologically and economically sustainable solutions in the tourist development, these resources and discourses should be examined and fitted together to find best possible compromises for the future development.

Some of the respondents felt that tourism development had changed their perception of their home village, and the once a peaceful village in the wilderness had become a tourist resort with snowmobiles and noise. For them, Kilpisjärvi was not as appealing as a place to live in as before, while others accepted the change as a necessity. It can be argued that the importance of the perceptional sphere of Kilpisjärvi, and the effect that these perceptions have on the attitudes toward tourism development in the village, has been underestimated and should be taken into account in the future planning of Kilpisjärvi. It should also be noted that tourism development and construction projects affect not only the residents' way of seeing the tourist destinations they live in but the tourists' perception as well, as Varley and Medway (2011) have discovered.

Reindeer herding was one of the discourses that contradicted tourism development most evidently. These two means of livelihood compete for the same areas, but there was also a more profound conflict between these discourses. Sámi reindeer herders have used the area for centuries, and for them, the meaning of the place was part of their culture. They also felt that their means of livelihood was treated unequally compared with tourism, with the latter bringing more economic benefits to the municipality.

The discourse of planning and administration and the discourse of possibilities to affect decision-making were also strongly addressed. Many conflicts in local tourism development discourses derive from poor administration and participation processes. The multiple role of the municipality in the land use planning, and partly also in implementation and supervision of the plans, seems to be problematic. There was also obvious mistrust between the administrators and many stakeholder groups. This mistrust and criticism were also reflected in the atmosphere of the village, complicating the planning process. As stated earlier, for instance, by Jokinen and Sippola (2007) and Tuulentie (2007), the participation process is a prerequisite for sustainable tourism. A participatory planning process enables collaboration, negotiation, compromises, and learning (Wray 2009) – outcomes that would be helpful in reconciling different discourses in a case like Kilpisjärvi.

As a general remark, we conclude that the discourses of tourism development in Kilpisjärvi have a strong social dimension, and we suggest that tourism development should be treated as a social web of discourses. Our intention is not to depreciate the importance of environmental issues in tourism development but rather to remind stakeholders that besides taking care of the environment, the diverse human interests should also be more essentially integrated into the planning of nature-based tourism destinations. Thus, it can be stated that in order to advance to a more sustainable situation, multiple voices must be heard in the local communities. The steps toward socially sustainable solutions in the planning process must include all stakeholders equally and must feature transparency and trust among all stakeholders and planners. After these steps, the commonly accepted goals and commitment to the process and the results can be achieved. While this task is not a simple one, the planning process itself can generate sustainability and empowerment and help to avoid a great loss of energy and money that would likely occur in difficult land use and development disputes.

References

Arnegger J, Woltering M, Job H (2010) Toward a product-based typology for nature-based tourism: a conceptual framework. J Sustain Tour 18(7):915–928

Berger PL, Luckmann T (1966) The social construction of reality. A treatise in the sociology of knowledge. Anchor Books, New York

Calhoun C (2011) Discourse analysis. In: Oxford reference online. Oxford University Press, Oxford. http://www.oxfordreference.com. 9 Nov 2011

Castellani V, Sala S (2010) Sustainable performance index for tourism policy development. Tour Manage 31:871–880
Choi HC, Murray I (2010) Resident attitudes toward sustainable community tourism. J Sustain Tour 18(4):575–594
Cooper C, Fletcher J, Fyall A, Gilbert D, Wanhill S (2005) Tourism. Principles and practice, 3rd edn. Pearson Education, Harlow
Cosgrove DE (1984) Social formation and symbolic landscape. The University of Wisconsin Press, Madison
Daniels S, Cosgrove D (1988) Introduction: iconography and landscape. In: Cosgrove D, Daniels S (eds) The iconography of landscape. Cambridge University Press, Cambridge, pp 1–8
Dredge D (2010) Place change and tourism development conflict: evaluating public interest. Tour Manage 31:104–112
Gunn CA, Var T (2002) Tourism planning: basics, concepts and cases, 4th edn. Routledge, London
Hall CM (2008) Tourism planning: policies, processes and relationships. Pearson Education, Harlow
Haukeland JV (2011) Tourism stakeholders' perceptions of national park management in Norway. J Sustain Tour 19(2):133–153
Heikkinen H, Jokinen M, Valkeapää OA, Helle T (2005) Poronhoidon historia Käsivarren Lapissa ja Mallalla. In: Jokinen M (ed) Poronhoidon ja suojelun vaikutukset Mallan luonnonpuistossa. Metsäntutkimuslaitoksen tiedonantoja 941. MetsaÌĹntutkimuslaitos, Kolari, pp 14–24
Itkonen TI (1948) Suomen lappalaiset vuoteen 1945. Ensimmäinen osa, 2. Painos. WSOY, Porvoo
Jokinen M, Sippola S (2007) Social sustainability at tourist destinations – local opinions on their development and future in Northern Finland. In: Jokimäki J, Kaisanlahti-Jokimäki ML, Tuulentie S, Laine K, Uusitalo M (eds) Environment, local society and sustainable tourism. Arctic Centre Reports 50. Arctic Centre, University of Lapland, Rovaniemi
Korpijaakko-Labba K (2000) Saamelaisten oikeusasemasta Suomessa – kehityksen pääpiirteet Ruotsin vallan lopulta itsenäisyyden ajan alkuun. Diedut No 1/1999. Pohjolan Painotuote Oy. Rovaniemi
Land Use and Building Act (1999) http://www.finlex.fi/fi/laki/kaannokset/1999/en19990132.pdf
Logar I (2010) Sustainable tourism management in Crikvenica, Croatia: an assessment of policy instruments. Tour Manage 31:125–135
Matless D (1998) Landscape and Englishness. Reaktion Books, London
McCool SF (2009) Constructing partnerships for protected area tourism planning in an era of change and messiness. J Sustain Tour 17(2):133–148
McDonald J (2009) Complexity science: an alternative world view for understanding sustainable tourism development. J Sustain Tour 17(4):455–471
Mettiäinen I, Uusitalo M, Rantala O (2009) Osallisuus matkailukeskusten suunnittelussa. In: Tuulentie S (ed) Turisti tulee kylään. Matkailukeskukset ja lappilainen arki. Minerva Kustannus, Helsinki
Plummer R, Fennell D (2009) Managing protected areas for sustainable tourism: prospects for adaptive co-management. J Sustain Tour 17(2):149–168
Puhakka R, Sarkki S, Cottrell SP, Siikamäki P (2009) Local discourses and international initiatives: sociocultural sustainability of tourism in Oulanka National Park, Finland. J Sustain Tour 17(5):529–549
Saarinen J (2007) Tourism in peripheries: the role of tourism in regional development in Northern Finland. In: Müller DK, Jansson B (eds) Tourism in peripheries. Perspectives from the far north and south. CABI International, Wallingford
Tuan YF (1974) Topophilia: a study of environmental perception, attitudes, and values. Columbia University Press, New York
Tuan YF (1977) Space and place: the perspective of experience. University of Minnesota Press, Minneapolis

Tuulentie S (2004) The power of nature-based tourism: tensions between different understanding of nature at the peripheral tourist resort of Kilpisjärvi. In: Saarinen J, Hall CM (eds) Nature-based tourism research in Finland: local contexts, global issues. Metsäntutkimuslaitoksen tiedonantoja 916. Finnish Forest Research Institute, Rovaniemi Research Station, Rovaniemi
Tuulentie S (2007) Local participation as a requisite for socially sustainable tourism: case studies from the Ylläs and Levi ski resort in northern Finland. In: Jokimäki J, Kaisanlahti-Jokimäki M-L, Tuulentie S, Laine K, Uusitalo M (eds) Environment, local society and sustainable tourism. Arctic Centre Reports 50. Arctic Centre, University of Lapland, Rovaniemi
Tuulentie S, Mettiäinen I (2007) Local participation in the evolution of ski resorts: the case of Ylläs and Levi in Finnish Lapland. For Snow Landsc Res 81:207–222
Varley P, Medway D (2011) Ecosophy and tourism: rethinking a mountain resort. Tour Manage 32:902–911
Wray M (2009) Policy communities, networks and issue cycles in tourism destination systems. J Sustain Tour 17(6):673–690

Chapter 12
Will "Free Entry into the Forest" Remain?

Argumentation Analysis of Finnish and Swedish Discussions on Everyman's Rights

Seija Tuulentie and Outi Rantala

Abstract In Europe, the public right of access is most widely applied in the Nordic countries. So-called everyman's rights give everyone the basic right to roam freely in the countryside, no matter who owns or occupies the land. What makes everyman's rights problematic is that they do not cover activities which damage the environment or disturb others, but the understanding of damage and especially disturbance is situational and depends on subjective valuation. There has been a lot of debate around everyman's rights over the past decade. Also, several surveys have been made for administrative purposes. By analyzing argumentation in recent discussion in the Internet, we evaluate in the article what are the most urgent issues at stake in Sweden and Finland. Our results show that the discussed themes vary from private ownership and national identity to commercialization and lifestyle. Especially berry pickers from Asian countries and the commercialization of recreational nature use have generated as well extreme nationalistic arguments as worried expressions of the ownership rights. A more moderate stance is to suggest negotiations between parties and to give more information in public. New devices to conceptualize the use of nature in a different way are slowly emerging but are not yet adapted to public common sense discussion.

Keywords Everyman's rights • Public access • Nature-based tourism • Land ownership • Argumentation

S. Tuulentie, Ph.D. (✉)
Finnish Forest Research Institute, PO Box 16, 96301 Rovaniemi, Finland
e-mail: seija.tuulentie@metla.fi

O. Rantala, Ph.D.
Multidimensional Tourism Institute, Faculty of Social Sciences, University of Lapland, PO Box 122, 96101 Rovaniemi, Finland
e-mail: outi.rantala@ulapland.fi

D.K. Müller et al. (eds.), *New Issues in Polar Tourism: Communities, Environments, Politics*, DOI 10.1007/978-94-007-5884-1_12,

12.1 Introduction

Nature-based tourism, recreation-related services, and berry picking by foreign workers are examples of new commercial land-use forms that tempt to question the tradition of everyman's rights in the Nordic countries. In a recent chat about everyman's rights, for example, one questioner wants to know if it is right that German enterprises come with all provisions, equipment, canoes, and buses with 40–50 persons and settle to a nice place. The answer by nature conservation jurist is that it is not clear: the best for all would be if the enterprise asks permission from the landowner. But this is a kind of tourism that did not exist decades ago and traditional everyman's rights do not recognize it (16SWE[1]).

In Europe, the public right of access is most widely applied in the Nordic countries. It is a basic element in the Scandinavian and Finnish outdoor life tradition and has traceable roots at least back to the county laws of the Middle Ages (Kaltenborn et al. 2001). The most important features of everyman's rights are that they give everyone the basic right to roam freely in the countryside, no matter who owns or occupies the land. Also camping for a short period is possible as well as picking berries and mushrooms (Finnish Ministry of the Environment 2007). These principles are simple and quite unambiguous. However, the statement concerning all everyman's rights is that they do not cover activities which damage the environment or disturb others is open to various interpretations. Damages and disturbance are situational and depend on subjective valuation.

Only Norway has a specific Open-Air Recreation Act (Friluftsloven). In Sweden, the right of public access is named as a concept in both the Environmental Code and the Swedish Constitution, but its content is not specified (Sandell and Fredman 2010). In Finland, everyman's rights are included in several acts but not collected into one specific act. However, it can be said that everyman's rights have evolved over the centuries from a largely unwritten code of practice to become a fundamental legal right (Finnish Ministry of the Environment 2007). During the last decades, changes in land-use modes have put more pressure to the interpretation of the rights. Some have felt that the rights are threatened, while some have demanded more restrictions to the rights.

National surveys in Finland (Viljanen and Rautiainen 2007; Lehtonen et al. 2007) and in Sweden (Sandell and Fredman 2010) have proved that everyman's rights are widely supported and accepted. However, attitudes in media and in other public fora in Finland are more critical than in survey replies (Lehtonen et al. 2007), and the debate around everyman's rights becomes extremely heated now and then. The same applies to the situation in Sweden. According to Sandell and Fredman (2010), the main problems arise from the organized commercial use which is seen as problematic in some local debates. Commercial use has been criticized by, for

[1]The reference (Number + SWE/FI) refers to our empirical data of blog-writings in Sweden (SWE) or Finland (FI).

example, the farmers association and the conservative party in Sweden. On the whole, increase in recreational and other kinds of nature use and new modes of use have also increased conflict potential around everyman's rights. This controversy becomes evident through the fact that the authorities regularly carry out clearings on the issue. In Finland, the Natural Resource Council (2008) stated after the survey reports in 2007 that the content of everyman's rights is sufficient and no alterations are needed, although the need for more information, active communication, and negotiations with commercial land use were recommended. In Sweden, the Swedish Environmental Protection Agency has nominated a committee to make an account of everyman's rights in order to get an idea of the future needs regarding the right codes (Naturvårdsverket 2011).

In this chapter, we ask what different forms the discussion around everyman's rights has taken in Finnish and Swedish national and local contexts and what those moral arguments that dominate the discussion of justice in each case are. We are also interested in the ways the discussion becomes polarized in the predominant argumentative culture. We compare the discussion in Sweden and Finland to each other, but we also look at the specific features that different issues inside everyman's right discussion get.

12.2 Data, Method, and Theoretical Starting Points

Our data consists of 40 blogs published in the Internet in Finland (20 writings; 44 printed pages with pictures and discussion) and Sweden (20 writings; 66 pages) concerning everyman's rights. They have been found through Google Blogs search engine with either the phrase "everyman's rights" (in Finnish: jokamiehenoikeudet; in Swedish: allemansrätten) or "everyman's rights and tourism." Thus, part of them concentrates on tourism, while others deal more generally with everyman's rights. The number of the writings taken for analysis was limited to 40 as they covered extensively the topics familiar from other media and public discussion, and no new themes arose with more writings. What were left out were the topics that handled everyman's rights more as a metaphor than as part of land-use practices. Thus, common to the selected writings is that everyman's rights are related to concrete land-use issues. The texts are fairly short: they vary from about 100 words to slightly over 1,000 words, Finnish writings being shorter than the Swedish ones. In some cases, the comments by the audience and answers by the original writer are so integral part of the content of the text that they are included in the data. Pictures can also be regarded as part of the argument. The nature of the writings varies from more traditional media commentaries to more informal Internet discussions.

In general, weblogs can be defined as "frequently published, date stamped entries in reverse chronological sequence" (Lehti 2011). In addition to this technical definition, a more socio-pragmatic perspective is needed: as Lomborg (2009) has stated, the blog is an author-driven, asynchronous, and informal genre of computer-mediated communication that uses various modalities and entails some interactivity (see also Lehti 2011). The nature of blog-writings is diverse, and the motivations of blogging vary a lot: some blogs are more official being part of the traditional media discussion, while some are more personal and informal, even spontaneous release of emotion (see, e.g., Nardi et al. 2004). Many of the blogs take the form of diary (Lehti 2011). Writers are either private persons, local or national politicians, or journalists. However, what is common to all blog-writings, and of interest here, is that they are public writings and almost all of our data can also be described as argumentative. The public sphere is an important field where social problems are constructed and political alternatives become defined (Koopmans and Statham 1999), and argumentativeness differentiates blogs, for example, from news articles.

Our data shows clearly that blogs are an integral part of political argumentation since many of the writers explicitly connect themselves to some political stance, and some others give an impression that they have party-political motives for writing. Argumentative stances are not usually adopted in isolation from wider issues. A number of studies have shown that when people debate a specific issue, the arguments they make are typically a small part of a much broader worldview, and, thus, conflict does not arise between isolated attitudes but rather attitudes embedded within entire worldviews (Plummer 2003). Quintessentially, argumentation takes polarized positions: social linguist Deborah Tannen has argued that much of discussion is dominated by "an adversarial frame of mind," where the best way to get anything done is to set about establishing, often quite aggressively, polarized positions (Tannen 1998; Plummer 2003). Binary thinking is typical in many kinds of situations, and explanations for, for example, environmental problems depend on the way of thinking behind the question. In the case of everyman's rights, for example, commercialization can be seen through the dualism of acceptance or denial of the constitutional rights to private ownership or in the sphere of all citizens possibility or impossibility to enjoy nature in relation to binary thinking that exists in several kinds of situations. The idea of contrast space (Garfinkel 1981) suggests that questions and answers arise from conflicting frameworks and thus have no possibility to communicate with each other (Haila 2002, 2004).

Sandell and Fredman (2010) have distinguished several possible conceptual frameworks in everyman's rights approach starting from environmental history perspective and environmental education and ending up to political identity and rights issues and the land-use perspective. Their own perspective derives from this latter point of view with emphasis on the theoretical framework of the "commons" and social capital. They also list three aspects which are relevant to the issue of recreational landscape's accessibility, namely, physical, legal, and sociopsychological. Social and psychological aspect comes closest to our perspective but not with such emphasis on individual dimension as Sandell and Fredman describe. According to them, social and psychological aspects "simply relate to the extent to which a specific landscape in a specific situation is experienced by a specific person as being accessible or not. This is about cultural background in a wider sense and here aspects such as upbringing, age, practices and knowledge become crucial" (Sandell and Fredman 2010, 293). However, Sandell and Fredman enlarge the view of social aspect by using the concept of social capital with reference to Ostrom. Ostrom

(2000) points out that social capital includes shared knowledge, understandings, norms, rules, and expectations about interaction that groups of individuals bring to a recurrent activity. Ostrom contends that social capital is formed over time and is embedded in common understanding rather than in physically obvious structures. This common understanding is difficult to articulate in words and is easily eroded if, for example, large numbers of people are concerned or if a large proportion of participants changes rapidly.

As difficult as it is to articulate this common understanding, it is to articulate practices and habits that are related to the use of nature. However, our purpose is to analyze how argumentation in the writings represents the changes in nature-related practices. Practices refer to the idea that human-nature relationships mainly consist of practical acts which are often routinized, embodied, and shared (Schatzki 2001; Jokinen 2002) and, thus, in a "normal" established situation also tacit. We assume that in a conflictual or contested situation, these practices become more explicit.

Here we will, firstly, distinguish the themes of discussion and the claims the writers are making. After that we discuss the argumentative features in the writings, and, before conclusions, we relate the argumentative strategies to wider argumentative stances and worldviews that represent changes in nature-based practices.

12.3 Discussion Themes in Finland and Sweden

When comparing the discussion around everyman's rights in Finland and Sweden, it is evident that on the one hand it has dealt with similar aspects, while, on the other hand, the topics that are raised differ in the two countries. Common to the discussion themes in both countries have been concerns about commercial berry picking and commercial tourism use and, in relation to these and other issues, but in more abstract level, the status of private landownership and owners rights. In these themes the basic underlying question is "who is entitled to benefit of 'our' nature?" Another common issue is to describe everyman's rights as an integral part of the national identity.

Contrary to Finland, much dispute in Sweden has raised of such concrete issues as building in the waterfronts and charging from the use of cross-country skiing tracks. In Finland, such acute specifically Finnish issues do not exist. However, in Finland, the rural landowners' organization is more present in the discussion, and the debate around foreigners' right to pick berries is more heated than in Sweden.

Topicality of everyman's rights in both countries is obvious. In Finland, the report by the Ministry of Environment begins by stating that "the use of everyman's rights is related to land ownership, to the constitutional protection of property, to the right of possession and to the domestic peace" (Viljanen and Rautiainen 2007, 3). The main reasons for the making of the report were expressed in the following way: "Everyman's rights are extremely important, and they are a topic of continuous discussion" (Viljanen and Rautiainen 2007, 3). Also the Finnish Rural Policy Committee launched its report making in 2006 because of "the wild

Table 12.1 The share of the writings related to the main themes. Note that some writings had several, equally notable themes and, thus, *N* is bigger than 40

The main theme of the writing country	Private ownership	National ownership and identity	Ecology	Commercialization	Lifestyle and culture	*N*
Sweden	9	8	2	12	9	40
Finland	4	9	–	9	10	32
N	13	17	2	21	19	72

stories from the field and from media which informed that the land owners have started to make restrictions while the nature entrepreneurs and urban nature-lovers have stretched the interpretation of everyman's rights to the utmost" (Lehtonen et al. 2007,1). Thus, the problems of everyman's rights have given the impulse to the accounts in both countries. Not surprisingly, the result of the latter study was that "the expert discussion in the media was more conflict oriented than the views expressed in the survey. Compared to the recent media coverage, the results of the survey showed more satisfaction with the functionality of the current everyman's right in all respondent groups" (Lehtonen et al. 2007, 59).

The argumentation in blog-writings follows and mirrors the public discussion that has created demand for the reports and accounts. In our data, we have distinguished between five argumentative themes that were most discussed. The themes are private ownership, national ownership and identity, ecology, commercialization, and lifestyle and culture. In every theme there exist contrasting claims. In the sphere of private ownership, the arguments are divided into ones that demand more authority and rights for the owners and into those that claim that private ownership restricts everyman's rights. National ownership and identity are highlighted especially with the emphasis on publicly owned forests and immemorial cultural rights. Ecology is not discussed in Finland at all, but in Sweden it is related to everyman's rights. Discussions about the themes of commercialization and lifestyle issues are the most diverse including topics such as new technologies used in leisure time and overall urbanization (Table 12.1).

National ownership and identity issues are especially in Sweden intertwined with lifestyle and culture issues. Swedish writings emphasize everyman's rights' uniqueness and their role as an integral part of national culture more than the Finnish ones. However, two of the writings remind that it is not a question of an immemorial right than often considered.

12.4 Argumentative Features of the Weblogs

A closer look at the different argumentation strategies shows that the Swedish writings express more strongly than the Finnish ones that everyman's rights are threatened ("hotad" in Swedish) or in danger ("i fara" in Swedish). Also, legitimacy and legality of new restrictions such as charges for skiing tracks is largely questioned. "Our Swedish culture," "our nature," and the uniqueness of Swedish everyman's rights are often used phrases. Argumentative use of "us" or "our" is typical for political creation of coalitions and in both expressing that we are all in the same boat but also in showing the speakers authority and, thus, strengthening national identity (Fairclough 1989; Tuulentie 2001). Emphasis on the difference between "us" and "them" also means that some people or groups are discursively constructed as outsiders. This is evident in a blog-writing where the writer states that everyman's rights are drummed into the souls of Swedish people, and, thus, it is easy to guide the Swedish (15SWE). However, according to the writer, the newcomers are a problem: they are not "normal tourists," he says, and they test all the rights. These people, according to the writer, are isolated from the society and do not get such information as, for example, tourists from tourist office.

Similarly, in Finnish writings, tourism and tourists are not such a big issue as are the foreign berry pickers who come with tourist visa but whose purpose is to make money and not to have vacation. They seem to form a kind of hybrid which is not easy to categorize. Especially in the Finnish writings, berry picking generates more strong and adversarial argumentation than other topics. Berry picking by foreigners is referred to as "berry war," and demands for conserving the "local immemorial picking places" are expressed.

The descriptions of the difference between berry pickers and tourists are interesting as they come close to Georg Simmel's classical sociological figure of the stranger whom Simmel defines as problematic since he is not "the wanderer who comes today and goes tomorrow" but rather a person who comes today and stays tomorrow (Simmel 1908/1971, 143). This feature distinguishes the stranger from the figure of tourist who is easy to categorize. According to Simmel, integral to a stranger is that he is "by his very nature no owner of land - land not only in physical sense but metaphorically" (Simmel 1908/1971, 144). However, for Simmel, being a stranger – a tradesperson – is a positive position, while in the context of everyman's rights, the attitudes of the locals are not always so positive. Still, in accord with Simmel's idea, the writings show that these more or less unclear positions are more problematic for the locals than "simple" tourist positions. This, we assume, is due to habitual positions and customary practices: tourists in charter buses or in other group formations are a familiar sight, but individual tourists wandering around are not – not to mention berry pickers from distant countries.

Another topic that is discussed with strong and adversarial rhetoric is that of commercialization. Critique to commercial berry picking is part of that discourse but the topic also involves charges for cross-country skiing tracks, waterfront constructions, and the use of nature by tourism enterprises. The strongest expressions such as putting the "entrance fee" to nature (8SWE) include connotations that confront forcefully against the Nordic idea of everyman's rights. Similarly, the danger of cutting down government expenses by putting entrance fees to national parks is regarded as the first step of making also other areas liable to charge (15FI). This is a typical rhetorical figure of presenting something as a symptom of larger phenomenon which can emerge if one concession is made (Fiske 1990; Tuulentie 2001).

Property owner's rights, socialism, and communism are examples of big ideological issues that are also often brought up in relation to everyman's rights. An example of an extremely colorful expression is the following: "Wallet freedom apologists are seldom advocates for real freedoms, they are mostly just Selfishness's temple masters" (20SWE). This argument is related to the idea that there is a problem with misuse of everyman's rights but ending them is no solution. Writer's solution is that those who misuse the rights should be punished, even if they were tourists.

Those who oppose the restrictions to everyman's rights question the possibility to own nature with argumentation that is derived from the philosophical ideas of man-nature relationship and democracy. For example, the question how to draw the line between nature and human territory (8FI) is brought up. Similarly, it is asked which part of land and natural resources can belong to the owner of the land: different kinds of restrictions to private ownership are needed in order to increase people's democratic rights (7SWE).

In addition to strong arguments, irony is also used as a rhetorical device: with the title "They come and take our bilberries" the writer ironizes the need for an established 'armchair service for berry picking' which he imagined to be for the mistreated foreign pickers but which turned out to be for the well-off locals who felt that their rights have been trampled (9FI).

The other side of the coin in commercialization discourse is that of the landowners. In Finland, the tone is instructive: the Central Union of Agricultural Producers and Forest Owners (MTK) has released a newsletter that has been used as a basis for blog-writings (12Fi; 13FI) and the tone is very moderate. The writings emphasize that everyone has to use common sense and good manners but the new modes to use nature are challenging. The problem from this perspective is seen as a lack of knowledge and understanding but also unconcern.

12.5 Argumentative Stances in Relation to Changes in Practical Use of Nature

Our analysis shows that in the discussion on everyman's rights, strong positions are taken and adversarial argumentation is used, especially in relation to commercial utilization of nature and private ownership. Usually argumentative stances are not adopted in isolation from wider issues. This means that conflict does not arise between isolated attitudes but rather attitudes embedded within entire worldviews (Plummer 2003). An issue that can be seen as a question of wider, often tacit worldview is the long Nordic tradition that recreation in nature is a common right and not related to land ownership. This basically simple idea is tested because of emerging new practices and worldwide political changes. Earlier there has not been any need to make a distinction between country's own citizens and the others. The need for this distinction awoke discussion already in relation to European Union in 1994 when there were concerns both in Sweden and Finland that the EU will deny everyman's rights. These fears were turned down by assuring that rights can remain as long as they are similar to all EU citizens (Westerlund 1991; Miettinen 1997). In southern Sweden the tourist groups who stayed by virtue of everyman's rights were regarded as a problem already in 1990s, but in Finland the problem was not apparent at that time (Miettinen 1997).

Nevertheless, it is not only a question of citizenship but also about the attitudes toward those who are more problematic strangers: the tourists are not such a problem as the people staying in the country without inherent understanding what is meant by everyman's rights (e.g., 15SWE). Thus, wider issues related to the berry picking theme seem to be related to overall attitudes toward "others" coming to "our" lands. Grounds for these arguments are derived from such nationalistic discourse as the historical right of the citizens to the nature-based products. In many ways, the concerns are related to the transitions in the uses of nature. On the one hand, the writers are worried that Finnish or Swedish people themselves, due to urbanization, do not have enough knowledge or understanding about everyman's rights and do not use and value nature in traditional ways. On the other hand, traditional habits are claimed to be threatened by the strangers, newcomers, and visitors, of different kinds.

Strong arguments and polarized position are part of what is called "the argument culture" which can be seen as an integral part of our culture on whole (Tannen 1998). Expressions, for example, with the word "war," are easily used as in the case of "berry wars." Confrontation in everyman's rights discussion can be seen somewhat strange as all parties are fairly unanimous that everyman's rights should be maintained largely in their existing form. The controversy, however, originates from the uncertainty of new land-use modes and practices. They are referred as everyman's rights, although they may not be included in the concept at all – as is in the case motorized off-road vehicles or reindeer herding – or their connection to everyman's rights is in any case strongly contested as is in the case of commercial utilization of private land.

Through the blog-writings, it becomes obvious that new tourism and other practices of nature use give more pressure to communicate and clarify the contents of everyman's rights. However, also more traditional ways to use the rights seem to be blurred: many writings demand that people should be informed and encouraged to use everyman's rights. This refers mainly to the fact that contemporary Nordic people are not so eager anymore to pick berries or camp in nature. Thus, this combined with the situation in which more and more foreigners and "outsiders" use "our" lands seems to cause confusion. An example of the testing of the limits of everyman's rights comes from the capital area of Finland where a blog-writer decides to live in a tent for four weeks while his apartment is renovated. He writes that he made an intriguing observation that people living in cities are not acquainted with everyman's rights:

> I have had the same discussion with several people: What are the places where you are allowed to camp and what are those where you are not? – It was surprising for me how many of us think that it is impossible to camp in the southern parts of Espoo municipality [close to the capital city of Helsinki] because of a lack of permitted places. – I even started to think that I'll have to call the police in advance and tell about my project as there will probably be people who call them about my tent. (14FI)

Thus, the need for the public discussion on everyman's rights does not only derive from the touristic and other commercial use but also from the situation that native Nordic people use nature more in a touristic way and are not familiar with the traditional ways of recreation in nature.

12.6 Conclusion

Blog-writings and other discussions around everyman's rights show that something both in attitudes and practices is changing – and has been for some time although the juridical position of everyman's rights has not changed. In relation to Finland's membership in European Union, there was a discussion on how the other Europeans can use Finnish nature with everyman's rights. Now, with the emergence of Asian and other berry pickers outside EU and more individual foreign tourists, it is becoming part of common knowledge that the users' nationality does not restrict their possibility to use nature according to everyman's rights. This fact generates as well extreme nationalistic arguments as worried expressions about the ownership rights. A more moderate stance is to suggest negotiations between parties and to give more information in public. The more extreme argumentation uses the rhetoric of "us" and "them" which can have more severe consequences – both in the practices of recreational use of nature and in the attitudes toward foreign workers and even tourists.

How do these different views communicate with each other? Do they have anything in common and is mutual understanding possible? The common ground is obviously in the consensus that everyman's rights are an important part of Nordic tradition and have to be maintained. However, shared understanding of what is harmful and disturbing use seems to be disappearing. Paying for the use of land for recreational purposes still evokes strong feelings but also the modes of behaving in nature create criticism.

New devices to conceptualize the use of nature in a different way are slowly emerging but are not yet adapted to public common sense discussion. In the fields of social sciences, and even more so in the administrational realm, the concept of ecosystem services (Millennium Ecosystem Assessment 2005) is one that is increasingly used when trying to solve the wider problem between different uses of natural resources, including recreational use, but the contents of the concept are still fairly unclear. However, in any case wider societal discussion about the possibilities and limits of recreation in nature is needed to make new issues more accessible and the strangers more familiar.

References

Fairclough N (1989) Language and power. Longman, London

Finnish Ministry of the Environment (2007) Everyman's rights in Finland: public access to the countryside: rights and responsibilities. Edita Prima, Helsinki. Available at: http://www.ymparisto.fi/download.asp?contentid=25603&lan=fi. Accessed 27 Oct 2011

Fiske J (1990) Introduction to communication studies. Routledge, London

Garfinkel A (1981) Forms of explanation: rethinking the questions in social theory. Yale University Press, New Haven

Haila Y (2002) Scaling environmental issues: problems and paradoxes. Landsc Urban Plan 61:59–69

Haila Y (2004) Retkeilyn rikkaus. Luonto ympäristöhuolen aikakaudella. Kustannus Oy Taide, Helsinki

Jokinen A (2002) Metsänomistaja metsänsä hoitajana. Rutiinit, "tarjokkeet" ja vastavuoroisuus. Yhteiskuntapolitiikka 67(2):134–147

Kaltenborn BP, Haaland H, Sandell K (2001) The public right of access – some challenges to sustainable tourism development in Scandinavia. J Sustain Tour 9(5):417–433

Koopmans R, Statham P (1999) Political claims analysis: integrating protest event and political discourse approaches. Mobilization 4(2):203–221

Lehti L (2011) Blogging politics in various ways: a typology of French politicians' blogs. J Pragmat 43:1610–1627

Lehtonen S, Heikkinen T, Hirvonen J (2007) Jokamiehenoikeuksien tulkintoja ja haasteita luontoliikunnan ja -matkailun kannalta. Maaseutupolitiikan yhteistyöryhmän julkaisu 1/2007. Available at: http://www.maaseutupolitiikka.fi/files/50/YTR_1_07.pdf. Accessed 22 Aug 2011

Lomborg S (2009) Navigating the blogosphere: towards a genre-based typology of weblogs. First Monday 14(5). http://firstmonday.org/htbin/cgiwrap/bin/ojs/index.php/fm/article/view/2329/2178. Accessed 14 Nov 2012

Miettinen M-K (1997) Maanomistajien näkemykset jokamiehen oikeuksista ja niiden toimivuudesta. Master's thesis in Sport sociology, University of Jyväskylä, Jyväskylä

Millennium Ecosystem Assessment (2005) Available: http://www.maweb.org/en/index.aspx. Accessed 22 Aug 2011

Nardi BA, Schiano DJ, Gumbrecht M, Swartz L (2004) Commun ACM 47(12):41–46

Naturvårdsverket (2011) Naturvårdsverket gör utredning om allemansrätten. Available at: http://www.naturvardsverket.se/sv/Start/Friluftsliv/Allemansratten/Arbete-med-Allemansratten/Naturvardsverket-gor-utredning-om-allemansratten. Accessed 22 Aug 2011

Natural Resource Council (2008) Statement of the utilization of everyman's rights and intensification of communication. Available at: http://www.mmm.fi/attachments/ymparisto/luonnonvara/5yFmNCu5z/LVNn_kannanotto_jokamiehenoikeuksien_hyodyntamisesta_12.6.2008.pdf. Accessed 22 Aug 2011

Ostrom E (2000) Social capital: a fad or a fundamental concept? In: Dasgupta P, Serageldin I (eds) Social capital: a multifaceted perspective. The World Bank, Washington, pp 172–214

Plummer K (2003) Intimate citizenship: private decisions and public dialogues. McGill-Queen's University Press, Montreal

Sandell K, Fredman P (2010) The right of public access – opportunity or obstacle for nature tourism in Sweden? Scand J Hosp Tour 10(3):291–309

Schatzki T (2001) Introduction: practice theory. In: Schatzki T, Knorr Cetina K, von Savigny E (eds) The practice turn in contemporary theory. Routledge, London, pp 10–23

Simmel G (1908/1971) The stranger. In: Simmel G (ed) On individuality and social forms: selected writings. University of Chicago Press, Chicago, pp 143–149

Tannen D (1998) The argument culture. Moving from debate to dialogue. Random House, New York

Tuulentie S (2001) Meidän vähemmistömme. Valtaväestön retoriikat saamelaisten oikeuksista käydyissä keskusteluissa. Suomalaisen Kirjallisuuden Seura, Helsinki

Viljanen P, Rautiainen A (2007) Jokamiehenoikeuden toimivuus. Report by the Ministry of the Environment 19/2007. Available at: http://www.ymparisto.fi/download.asp?contentid=74190. Accessed 22 Aug 2011

Westerlund S (1991) EG:s miljöregler ur svenskt perspektiv. NaturskyddsfoÌĹreningen, Stockholm

Chapter 13
National Parks for Tourism Development in Sub-Arctic Areas: Curse or Blessing?

The Case of a Proposed National Park in Northern Sweden

Dieter K. Müller

Abstract National parks in peripheral areas are often promoted as tool for regional development and a way of restructuring resource-based economies into tourism. Though not always embraced by local communities, tourism businesses are generally expected to be in favor of new park establishments. However, the coalition between environmentalists and tourism entrepreneurs has recently been challenged. The purpose of this chapter is to analyze discourses regarding the establishment of national parks in Arctic areas. This is done regarding the proposed Vindelfjällen National Park in northern Sweden. Using discourse analysis of media debate, public documents, and interviews, it is demonstrated that struggle over national parks is not only about tourism development but also wider concern regarding the management of Arctic environments. In this context it is not possible to simply classify stakeholders in local, national, and international. Interests differ in both arenas and mirror rather a general struggle over control of northern areas. Hence, tourism development is used just as an arena for negotiating issues of governance and power.

Keywords Nature-based tourism • National parks • Core-periphery relations • Sweden • Regional development

13.1 Introduction

National parks[1] in peripheral areas are often promoted as tools for regional development (Eagles and McCool 2002; Goodwin 2000; Hammer 2007; Lundmark and Stjernström 2009). Already upon the creation of the first national parks in the

[1] Swedish national parks are classified as parks according to the IUCN category II. The discussion in this chapter refers to this type of protected area, mainly.

D.K. Müller, Ph.D. (✉)
Department of Geography and Economic History, Umeå University, SE-90781 Umeå, Sweden
e-mail: dieter.muller@geography.umu.se

D.K. Müller et al. (eds.), *New Issues in Polar Tourism: Communities, Environments, Politics*, DOI 10.1007/978-94-007-5884-1_13,

world, the railway and associated economic interests were identified as important factors in the establishment of parks (Frost and Hall 2009; Wall Reinius 2009). This demonstrates that positive regional impacts are usually expected to occur in relation to tourism development, and thus "worthless" land is turned into an economic asset not least for the benefit of local communities (Hall and Frost 2009). This is particularly true for peripheral areas, where nature is often abundant and in fact even consists of cultural sites unthreatened by alternative development (Hall 2007). It is thus remote areas that have often been selected for the establishment of national parks. Particularly Arctic and sub-Arctic regions have been of interest, not only for bringing economic benefits to peripheral regions but also for manifesting the sovereignty of remote territories (Hall and Johnston 1995; Lemelin and Johnston 2008; Timothy 2010).

However, tourism development in these northern regions is not always embraced or successful. Particularly distance to demand markets and a lack of product supplies are major constraints for development (Hall and Boyd 2005; Hall 2007; Müller and Jansson 2007). Moreover, Sandell (1995) reports that local resistance in northern Sweden indeed stopped a proposed national park. He argues that this was mainly due to a local fear of the national park designation creating restrictions on outdoor recreation and resource use for hunting and fishing. Similarly, Kaltenborn et al. (2011) and Puhakka et al. (2009) report on dissatisfaction with comanagement practices of national parks in Norway and Finland, respectively. Their arguments are in line with Lundmark and Stjernström's (2009) concern that national park designations in fact transfer power over land use from the periphery to core regions. Moreover, Lundmark et al. (2010) demonstrated that national parks in northern areas do not have a guaranteed positive impact on local tourism labor markets, and hence, they claim that the positive relationship between nature conservation and tourism is at least questionable.

In Sweden the failure to establish a national park in the Kiruna mountains entailed a new approach concerning national parks whereby local interests have to be involved already at the beginning of the new park designation process. This was successfully done in the case of Fulufjäll National Park, where increased tourism was promoted as a major advantage of a national park designation. And indeed, the establishment of the national park increased visitation and tourism expenditure in the rather remote area (Fredman et al. 2007). However, when the same concept was applied in the proposed establishment of a national park in the Vindelfjällen area it failed, seemingly due to the resistance of tourist entrepreneurs, who are actually intended to be the group that benefit most from a national park.

Against this background, the purpose of this chapter is to analyze discourses regarding the establishment of national parks in Arctic areas. The case studied is the proposed Vindelfjällen National Park in northwestern Sweden. To achieve this, the contents of newspaper articles covering the process of negotiating the park are analyzed.

13.2 Tourism and National Parks in Northern Sweden

Traditionally, national parks are located in areas considered remote and marginal (Butler and Boyd 2000; Frost and Hall 2009). This is partly because of the ease of disposing land for the purpose of national parks and partly because of the presence of natural amenities. For the northernmost hemisphere, this means that parks have been predominantly located in sub-Arctic and Arctic areas lacking larger populations and infrastructures. Tourism has played an important role in this context, providing arguments and an alternative value for protecting land from exploitation. It has also given an additional value to infrastructure investments needed to explore northern natural resources. This is particularly true for the European North, where railway lines and road networks opened up the northern tourism space early on (Lundgren 1995; Wall Reinius 2009). Although this can be seen in the Canadian North too, here the availability of air transportation played a more significant role (Lundgren 1995, 2001). Today, growth is mainly in cruise tourism (Stewart and Draper 2006).

As Mels (1999) argues, national parks have to be seen in the historical context in which they have been created and developed. It is thus important to point out that Sweden's North, formerly seen as a land of the future owing to its richness of natural resources (Sörlin 1988), is now often perceived as a backwater area (Eriksson 2010). This image includes depopulation, an ageing society, lack of human capital and competence, and great economic dependence on outside support. Thus, the establishment of national parks today has to be seen against this background and as a remedy with the potential to at least fight these symptoms.

Certainly, the concept of and ambitions with national parks have changed since the early establishments, owing not least to the steady increase of protected areas and the subsequent need to justify this development (Mose and Weixelbaumer 2007). Regarding the Swedish case, Fredman and Sandell (2010) argue that there were four periods of outdoor recreation that influenced the establishment of protected areas and more or less followed international trends. A first phase can be traced back to the early years of the twentieth century, which saw the establishment of the first national parks in Europe, among them Abisko and Stora Sjöfallet (1909), both located north of the Arctic Circle. The argument for this protection was a mixture of nationalist ideologies: an emerging awareness of environmental change and a growing interest for outdoor recreation and tourism. A second phase started the 1930s, when tourism and outdoor recreation were clearly identified as integrated parts of the evolving Swedish welfare state and were promoted by laws and regulations, including the recognition of the Public Right of Access. This did not, however, lead to the establishment of new parks as the focus increasingly shifted to areas closer to major population centers. After the Second World War, a rapid expansion of outdoor recreation and nature-based tourism occurred, also leading to an increase in protected areas (Fredman and Sandell 2010). National planning identified areas of national interest for outdoor recreation, and numerous new nature reserves were also established. At the same time, the government maintained hiking tracks originally designed by the Swedish Tourist Association in the Swedish mountain range (Fredman and Sandell 2010).

At the end of the twentieth century, national parks again gained attention. An increasing environmental awareness – not least after the Brundtland Commission's Report, marking what could be called a sustainability turn in political rhetoric, as well as EU guidelines on protection – caused the consideration of new national parks, and the Swedish environmental protection agency (SEPA) was ordered to revise its national park plan. In this context, in 2000 the Swedish parliament decided on 16 environmental objectives highlighting the desired state of the Swedish environment in 2020 (Miljödepartementet 2000). An important objective is the preservation of a magnificent mountain landscape, and identified threats include contested biodiversity and high noise levels from human activity. As major means to achieve the objectives, the protection of large areas and particularly the extension of "quiet" areas were suggested.

The acceptance of national parks is not self-evident, however (Sandell 1995, 2007). Sandell (2007: 145ff) offers a conceptual framework of eco-strategies to discuss different human perspectives on nature. Accordingly, a major division is between "functional specialization" and "territorial adaptation." In the latter, the landscape is a given entity offering resources to be used or preserved. Sandell (2007) labels the related eco-strategy as one's home district. Even within this home district, use of landscape can vary between passive admiration and active utilization through hunting, fishing, or limited logging for local use. Functional specialization, too, can be divided into an active and a passive variant. The active dimension includes an improvement of the landscape to fit the proposed functional specialization, for example, changes necessary for the development of alpine ski resorts. Sandell labels this a factory perspective, aiming at providing activities. In contrast, a passive use puts the preservation of landscape in focus. Even here landscape is built, but by banning other activities. Sandell calls this a museum for external consumption. In this context tourism is entirely in the realm of functional specialization, while outdoor recreation highlighting the leisure practices of the local population is part of territorial adaptation too.

Sandell (2007) claims that national parks are historically an expression of a museum strategy, a process sometimes also labeled "embarkment" (Hermer 2002). However, the recent shift in conservation policies in Sweden and elsewhere, acknowledging the socioeconomic benefits of outdoor recreation and striving for greater economic revenue related to tourism, extends the national park concept into the realm of the factory strategy. Moreover, it can be argued that the increasing call for comanagement, including a variety of stakeholders in governance of protected areas and evolving not least as a reaction to the nonestablishment of national parks (Zackrisson et al. 2006), is an extension of the national park perspective in the direction of one's home district.

The extension of the national park concept in a direction that promises economic revenues from designation owing to greater tourism is doubtful, however. As Lundmark et al. (2010) have noted, national parks do not necessarily deliver the desired impacts, though among actual visitors designation seems to play at least some role (Wall Reinius and Fredman 2007). A national survey has revealed that only 6.5% of the Swedish population has visited a mountain national park (Fredman and Sandell 2010). Among residents of the mountains, the likelihood to visit a national park is greater, and they also have distinctively diverging attitudes toward national parks. Accordingly, local populations should have a greater influence in park matters, and more than 15% of the respondents residing in the mountains in fact feel there should not be extended protection in the mountain area (in contrast to 3.4% in Sweden in general).

This indicates that national parks in the northern periphery are mainly there to satisfy interests outside the area. Hence, it can be argued that a core-periphery dimension can complement Sandell's framework (2007). The museum strategy is more frequently apparent in core areas of southern Sweden, while the home district strategy is mainly visible in areas adjacent to the mountain parks. This will now be discussed with reference to the proposed Vindelfjällen National Park.

13.3 Background and Methods

The first time a newspaper reported on the proposal to establish a national park in Vindelfjällen was May 1989. The SEPA presented a new national park plan to be implemented during the 1990s and included Vindelfjällen as a potential national park due to its wilderness characteristics and mountain landscape. In fact, the area had already been assigned national nature reserve status in 1974 which was later extended, and thus special rules aiming at protecting the area were already in place. Altogether more than 500,000 ha are under protection, and all land is owned by the state (Fig. 13.1). However, reindeer herding and various outdoor recreation activities are allowed. Moreover, the nature reserve contains parts of the King's Trail, the major long-distance hiking trail through the Swedish mountains. There are no settlements within the area. Huts belonging to the Swedish Tourist Association are the only major service infrastructure in the reserve. Sami reindeer herders have cottages, too, allowing them to stay in the mountains during the summer grazing season.

Three small settlements are located on the outskirts of the reserve. Ammarnäs, with roughly 100 inhabitants, is located in Sorsele municipality and is surrounded by the nature reserve. Besides nature-based tourism, reindeer herding is another source of income for the local population. On the southwest border of the reserve in Storuman municipality are the villages of Tärnaby and Hemavan. These villages are alpine centers of domestic significance, and thus their 480 and 220 inhabitants, respectively, are accompanied by seasonal tourism workers at least during the winter season. Tärnaby also functions as a service center for the upper areas of the municipality, and indeed the area has had a balanced population development, not least thanks to amenity-led migration (Müller 2005). Moreover, an airport facilitates daily connections to Stockholm. For all three villages, second homes entail an addition to the registered population.

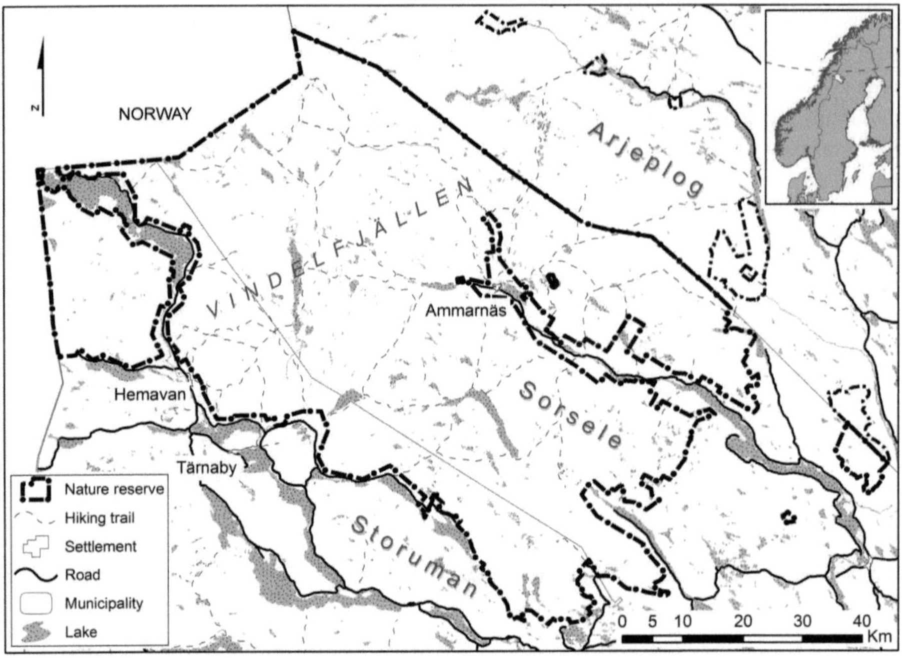

Fig. 13.1 The Vindelfjällen Nature Reserve

The tourism business in Ammarnäs is composed of seven different accommodation providers and a number of companies offering activities including fishing and horseback riding. In contrast, Tärnaby and Hemavan have more than 20 different providers of accommodation and tourist services. A major national tourism company has invested in the alpine skiing facilities and the hotel infrastructure (Bodén and Rosenberg 2004).

The initial national park plan from 1989, identifying Vindelfjällen as a potential site for a national park, was only partly implemented during the 1990s. It was not until 2007 that a draft for a revised plan was presented, again featuring the area as potential site for a park. In this context it was stated that national parks would not be established against the will of the local municipalities, which was certainly a lesson learned from earlier failures, and thus a pre-study was proposed to determine whether a national park would be feasible. This initiated debate in the two involved municipalities: Storuman and Sorsele. In fact, a third municipality to the north, Arjeplog, was initially involved in order to discuss an extension of the proposed park. However, this was directly rejected by the local government. In contrast, the other municipalities engaged in a discussion of whether they should agree on a pre-study evaluating the potential benefits of a park. In Storuman a majority rejected the park plans, though decisions were revised several times, while the situation in Sorsele was more complex, with shifting majorities and decisions in favor of and against a pre-study. Roughly a year after the presentation of the draft

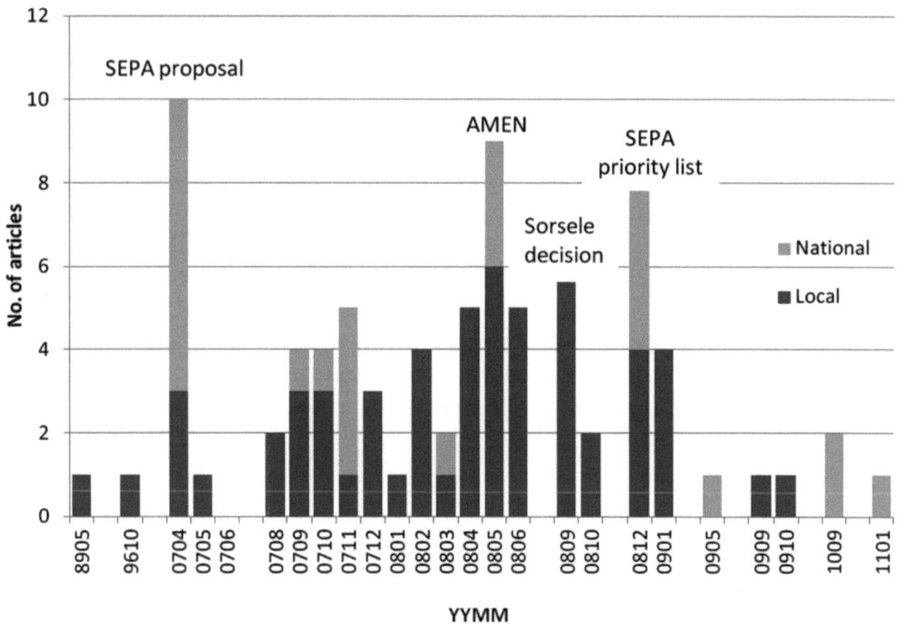

Fig. 13.2 Newspaper articles mentioning the proposed Vindelfjällen National Park

plan, the Action Group against a National Park (AMEN) was founded in Sorsele, clearly altering public opinion against a national park pre-study. In September 2008, Sorsele decided to reject a pre-study. Three months later the SEPA presented the revised plan including a priority list, which also entailed a discussion relevant to the establishment of a national park in the area. Moreover, in 2009 the Swedish National Park System celebrated its 100th anniversary, which also entailed a number of articles mentioning the proposed national park in Vindelfjällen.

The attempt to establish a national park is analyzed here utilizing a qualitative and thematic analysis (Boyatzis 1998) of Swedish newspaper articles on the issue. Thematic analysis is understood as a way of systematically observing interactions and groups. Here, the focus is on the content of newspapers that reflects the public debate about the national park. The material includes both editorial texts and opinions; thus the analysis concerns not only the content itself but also who said what. Thus, in the context of this chapter, an important question is where the articles originate. Do they reflect a local debate or national coverage of a local debate?

The sample of newspaper articles selected for this study was gathered from Mediarkivet, a full-text database of Swedish newspapers. Since Västerbottens Folkblad (VF), the second major local newspaper after Västerbottens Kuriren (VK), was not included in the database, VF's online archives were also searched. The keywords used for searching articles were "Vindelfjällen*" and "national park." This procedure identified a total of 88 articles (Fig. 13.2). During a first reading of the articles, a number of themes were identified and coded. Consecutively, a number of articles were deselected as they only repeated material from press agencies. Moreover, the number of categories was reduced, ordered according to a timeline of development mirroring the different phases of the process, and analyzed in depth. Finally, the results were completed using a set of documents provided by state authorities to supplement the background story of this chapter.

A limitation to the study is that no specific indigenous perspective was applied. This would be a useful complement acknowledging the special land rights of the Sami population in the area. However, a thorough analysis of indigenous attitudes toward a national park would exceed the scope of this chapter. However, the indigenous dimensions of current attempts to nature protection were highlighted by Green (2009) concerning the implementation of a world heritage area in Northern Sweden. In her study it is shown that nature protection has to be seen in the context of ethno-political struggle reaching far beyond the actual protection issue.

13.4 Themes

13.4.1 Size Matters for Development

An early theme after the launching of the revised national park draft was related to potential tourism development in relation to a national park designation. The SEPA promoted the national park status as a "quality label" and also as a state commitment to engage in the area. Accordingly, a SEPA officer is cited as stating

> These are areas that we love to see as national parks since they have high natural values and are great sights. [. . .] Establishing a national park implies a long-term state commitment to preserve, manage and show an area. It implies investments to host visitors and greater security for investments in nature-based and other tourism . . . (VK 2007-04-27)

This perspective was also later confirmed by the director of environment at the county council, who cited examples in which national parks motivated investments in new roads and campgrounds (VK 2007-06-26). Moreover, it was mentioned that Vindelfjällen would become Europe's largest national park, further elevating the status of the attraction. Another argument related to previous experiences of municipalities is affected by earlier establishments of national parks. In a SEPA survey, these municipalities were said to be happy with the outcome of the national parks' establishment (VK 2007-04-28).

A majority of the articles mirror a balanced debate in which several stakeholders argue for a pre-study simply to evaluate whether a national park would be feasible and beneficial for the local community (VK 2007-09-12). Thus, the population in the municipalities was split regarding their support for and opposition to the pre-study (VF 2008-02-13).

The municipalities' threat to withdraw from the pre-study was misunderstood and heavily criticized by nonlocal proponents of the park project (VK 2008-04-28). Indeed, the proposed national park had been labeled as a "jewel" among the parks

(VK 2008-06-25). Following the rejection of the proposed pre-study, a commentator noted how another missed opportunity to upgrade the area's road infrastructure had been missed (VK 2009-09-11).

13.4.2 A Park Against Mining

The nature reserve had been the site of explorative drilling for mining resources for some time (Müller 2011), which had resulted in a Canadian company expressing interest in starting a copper and zinc mine in the area. Representatives of the county council and the Mining Inspectorate of Sweden dismissed these aspirations as wishful thinking (Hallands Nyheter 2007-09-07). In fact the county council even expressed irritation with the Mining Inspectorate, accusing them of unnecessarily licensing explorative drilling (VK 2007-10-18, 2007-12-07).

The indigenous perspective was only covered briefly in the media. However, it was mentioned that the Sami parliament regarded the park initiative, including co-management, as part of a decolonization process of the Sami area and thus a way to secure Sami influence in management of natural resources in the area (VK 2007-12-28). This obviously included mining (VF 2008-02-19; VK 2008-02-20). Reindeer husbandry, which promised the SEPA, would not be affected at all by the establishment of a national park (VK 2008-09-11).

13.4.3 A "Dead Hand"

An immediate reaction to the draft plan concerned the local population's opportunities to use the area for outdoor recreation. The SEPA and county council, however, calmed these interpellations. A SEPA officer is quoted as follows:

> We won't lay a dead hand but a green hand over the area. There will be possibilities to drive snowmobiles, fish and hunt in the area, but exactly where and in what scope will be a question to be discussed with the municipalities, among others... (VK 2007-04-28)

Even the director of environment at the county council strengthened this perspective and stated that he could not imagine a ban on hunting. The establishment of a park against the will of the municipalities was more or less out of the question (VK 2007-06-26). Moreover, as the proposed park also included an extension into hitherto unprotected land, land owners were concerned about the potential dispossession of private land, which was obviously already a well-known phenomenon in the context of national park establishment (VK 2007-04-28).

However, the official arguments for the establishment of a national park did not impress everyone in the involved municipalities. Obviously, new restrictions on the use of the area were expected (VK 2007-08-15, 2008-03-13). There were also some voices that saw hinders to future exploitations of and economic benefits from natural resources in the area (VK 2008-09-23).

13.4.4 Mistrust in Core-Periphery Relations

A first indication of mistrust regarding the state and the SEPA in particular was noted as early as 2008. During discussions on the pre-study in Sorsele, concerns were raised regarding the trustworthiness of the counterparts:

> The decision [to establish the nature reserve] 1974 was made against the municipality's will and a general understanding is that promises were not fulfilled. (VK 2007-08-15)

AMEN took a similar stand, pointing out the limited economic impact of previously established parks. Moreover, they also highlighted the fact that the promise that a national park would not be established against the will of the local municipalities was not anchored in law and was thus not trustworthy (VK 2008-04-28), which turned out to be an important argument for the final "no" vote by Sorsele (VK 2008-05-17). This mistrust was further fuelled by the insight that Iceland was in fact planning a national park even larger than Vindelfjällen (VK 2007-08-23). This information was not disseminated to the local arena, however, although a SEPA officer admitted that he had been aware of this and other large-scale plans in Russia.

Another dimension was introduced by former competitive alpine skier Stig Strand, now an influential tourist entrepreneur in Hemavan. He mentioned the core-periphery problem related to national parks:

> A national park is governed by parliament and government. Reserves are governed by the county administration. There will be longer ways for getting exempts from bans. [...] Touristically there are absolutely no advantages. (VK 2007-09-11)

> There's no chance to survive here. If there's going to be a national park, I want to know this early on in order to be able to sell my business. Otherwise, we'll sit here and go bankrupt. (Dagens Nyheter 2008-05-05)

A consequence was a unanimous "no" vote by Storuman municipality on the pre-study. A leading politician is cited as follows:

> We simply do not trust the ambitions of the authorities in this case. It's happened so often that we've felt brushed aside by the authorities, who promised the moon and stars, but in the end it was just lies. (VK 2007-09-13)

Local politicians in the area north of the reserve also strongly rejected the plan and pointed at the risk that local comments on national policies were usually ignored (Dagens Samhälle 2007-10-25) and that the North had already contributed enough to national protection measures (Dagens Samhälle 2008-03-27).

However, there were also signs of a counter-discourse. A leading politician in Storuman mentioned the risk of losing decision-making power regarding the central state by saying no to a pre-study (VF 2008-02-14). Even tourism entrepreneurs in Ammarnäs, often members of the indigenous population, continued to see a national park as an opportunity (VF 2008-06-24; VK 2008-09-23, 2008-10-04). Moreover, the county administration said they would keep their promise and respect the votes of the local municipalities (VK 2008-05-31).

13.4.5 The Foundations of the Proposed Park

A single article highlighted the foundations on which the current park proposal was built. As a reaction to the debate that, according to the author of this opinion article, questioned the ability of the local population to make a reasonable assessment, he stated the following:

> Who is it that contributed to the fact that we have an almost untouched nature in Vindelfjällen? Yes, the Sami and the local population. We had the right people running tourism businesses [...] We have experiences prior to the times of colonization, and after state expropriation, of how people have been discriminated against, questioned and depreciated as village idiots. (VK 2008-04-21)

13.5 Epilogue

After the municipalities' dismissal of the national park pre-study, the SEPA presented the final version of the revised national park plan in December 2008. It included Vindelfjällen as a part of a more long-term plan (VF 2008-12-03), not least because there was a hope that local opinions would change in the future (VK 2009-12-04). Such disregard of the local vote also created unease in the adjacent southern municipalities, where the municipalities had agreed on a pre-study for another proposed national park (VK 2009-01-15, 2009-01-17).

In the national media, on the other hand, the decision against the pre-study was met with agony considering the noise intrusion from snowmobiles as something to cope with until the area finally became a national park (*Göteborgs Posten* 2011-01-28). In *Dagens Nyheter*, the leading Swedish daily newspaper, an editorial addressing the regionalization of nature preservation argued for in the national park plan put it more radically, depicting the transfer of the management of national parks to the regions as follows:

> It's as if the government would allow speeding offenders to be part of the decision-making regarding speed limits on our roads [...]. There's of course nothing that says that local actors [...] would be worse in safeguarding great nature values. But experience [...] unfortunately shows that this is the case. [...] It's not difficult to imagine that a municipal politician in a pressed economic reality has difficulty defending himself against ideas of investing in large-scale snowmobile tourism or anything else that collides with interests in nature. (*DN* 2010-04-26)

13.6 Conclusion

In 2010, even the SEPA realized that "Attaining the environmental objectives is frustrated by the fact that many stakeholders perceive a lack of respect for local perspectives from central/regional/local authorities/civil servants. The locally perceived legitimacy of authorities is often low owing to previous actions and experiences" (Naturvårdsverket 2010:12). However, as the last quote in the result section perhaps indicates, local populations' resentment of state initiatives in nature protection is not unmotivated. In fact, Lundmark and Stjernström's (2009) claims that protected areas function as eco-protectorates excluding local populations from influencing land use seem to be reaffirmed.

The results also indicate a path dependency in skepticism toward state agencies. Historical experiences are frequently used to reject current initiatives. And indeed, the experiences from this case certainly reaffirm this procedure as a correct response. Even this time the state promised and failed to keep its promises; the proposed park was included in the national park plan, breaking the promise to recognize the vote of the local population.

Further, the argument that national parks are good for tourism did not convince all tourist entrepreneurs. These findings support similar studies conducted by Kaltenborn et al. (2011) and Puhakka et al. (2009) and concerns voiced by local stakeholders regarding comanagement practices in national parks in Sweden.

Thus, the struggle for nature protection is greatly a core-periphery conflict, though significant variations occur in the respective regions; one result of this study is certainly the almost complete absence of the northern national park debate in the national media, mirroring the peripheral and marginal characteristics in the public perception (Eriksson 2010). With regard to Sandell's framework (2007), it can be argued that a majority of the national stakeholders seem to follow a functional specialization approach, even though an understanding of the local population's home district strategies is shown in the rhetoric. As a consequence it could be argued that tourism should not be placed, as suggested by Sandell (2007), entirely on the side of functional specialization in his framework. Instead, it is rather a question of whether stakeholders have an active or a passive approach to the landscape and their environment.

One comment found in the newspaper articles should be addressed specifically, however. If the current place characteristics, including the practices of local populations, have managed to create and sustain an area that is considered a pristine wilderness, why is there a need to protect the area from (or even for) exactly these stewards (VK 2008-04-21)? It is probably a matter of the dominating opinion in the Swedish core that northerners cannot be trusted when it comes to decisions that require competence and broader perspectives (DN 2010-04-26). With this core-periphery conflict, it is unlikely that the Swedish state will be able to establish further national parks in the North.

Acknowledgement The research presented in the chapter was financed by the Swedish Research Council (VR) and the MISTRA Arctic Future Program. The support is hereby gratefully acknowledged.

References

Bodén B, Rosenberg L (2004) Kommersiell turism och lokal samhällsutveckling: En studie av sex svenska fjälldestinationer. Etour, Östersund

Boyatzis RE (1998) Transforming qualitative information: thematic analysis and code development. Sage, Thousand Oaks

Butler RW, Boyd SW (eds) (2000) Tourism and national parks: issues and implications. Wiley, Chichester

Eagles PFJ, McCool SF (2002) Tourism in national parks and protected areas. CABI, Wallingford

Eriksson M (2010) (Re)producing a periphery: popular representations of the Swedish north. Department of Social and Economic Geography, Umeå

Fredman P, Sandell K (2010) Protect, preserve, present. In: Frost W, Hall CM (eds) Tourism and national parks: international perspectives on development, histories and change. Routledge, London, pp 197–207

Fredman P, Hörnsten Friberg L, Emmelin L (2007) Increased visitation from national park designation. Curr Issues Tour 10(1):87–95

Frost W, Hall CM (2009) Reinterpreting the creation myth: Yellowstone National Park. In: Frost W, Hall CM (eds) Tourism and national parks: international perspectives on development, histories and change. Routledge, London, pp 16–29

Goodwin H (2000) Tourism, national parks and partnerships. In: Butler RW, Boyd SW (eds) Tourism and national parks: issues and implications. Wiley, Chichester, pp 245–262

Green C (2009) Managing Laponia: a world heritage as arena for sami ethno-politics in Sweden. Department of Cultural Anthropology and Ethnology, Uppsala

Hall CM (2007) North–south perspectives on tourism, regional development and peripheral areas. In: Müller DK, Jansson B (eds) Tourism in peripheries: perspectives from the far North and South. CABI, Wallingford, pp 19–37

Hall CM, Boyd S (2005) Nature-based tourism in peripheral areas: introduction. In: Hall CM, Boyd S (eds) Nature-based tourism in peripheral areas: development or disaster? Channel View, Clevedon, pp 3–17

Hall CM, Frost W (2009) National parks and the 'worthless lands hypothesis' revisited. In: Frost W, Hall CM (eds) Tourism and national parks: international perspectives on development, histories and change. Routledge, London, pp 45–62

Hall CM, Johnston ME (1995) Introduction: pole to pole: tourism issues, impacts and the search for management regime in polar regions. In: Hall CM, Johnston ME (eds) Polar tourism: tourism in the Arctic and Antarctic regions. Wiley, Chichester, pp 1–26

Hammer T (2007) Protected areas and regional development: conflicts and opportunities. In: Mose I (ed) Protected areas and regional development in Europe: towards a new model for the 21st century. Ashgate, Aldershot, pp 21–36

Hermer J (2002) Regulating Eden: the nature of order in North American parks. Toronto University Press, Toronto

Kaltenborn BP, Qvenlid M, Nellemann C (2011) Local governance of national parks: the perception of tourism operators in Dovre-Sunndalsfjella National Park, Norway. Norsk Geografisk Tidsskrift 65(2):83–92

Lemelin RH, Johnston M (2008) Northern protected areas and parks. In: Dearden P, Rollins R (eds) Parks and protected areas in Canada: planning and management, 3rd edn. Oxford University Press, New York, pp 294–313

Lundgren JO (1995) The tourism space penetration processes in northern Canada and Scandinavia: a comparison. In: Hall CM, Johnston ME (eds) Polar tourism: tourism in the Arctic and Antarctic regions. Wiley, Chichester, pp 43–62

Lundgren JOJ (2001) Canadian tourism going north: an overview with comparative Scandinavian perspectives. In: Sahlberg B (ed) Going north: peripheral tourism in Canada and Sweden. Etour, Östersund, pp 13–46

Lundmark L, Stjernström O (2009) Environmental protection: an instrument for regional development? National ambitions versus local realities in the case of tourism. Scand J Hosp Tour 9(4):387–405

Lundmark L, Fredman P, Sandell K (2010) National parks and protected areas and the role for employment in tourism and forest sectors: a Swedish case. Ecol Soc 15(1):19

Mels T (1999) Wild landscapes: the cultural nature of Swedish national parks. Lund University Press, Lund

Miljödepartementet (2000) Framtidens miljö – allas vårt ansvar. SOU 2000:52. Miljödepartementet, Stockholm

Mose I, Weixelbaumer N (2007) A new paradigm for protected areas in Europe? In: Mose I (ed) Protected areas and regional development in Europe: towards a new model for the 21st century. Ashgate, Aldershot, pp 3–20

Müller DK (2005) Amenity migration and tourism development in the Tärna Mountains, Sweden. In: Moss LAG (ed) The amenity migrants: seeking and sustaining mountains and their cultures. CABI, Wallingford, pp 245–258

Müller DK (2011) Tourism development in Europe's "last wilderness": an assessment of nature-based tourism in Swedish Lapland. In: Grenier AA, Müller DK (eds) Polar tourism: a tool for regional development. Presse de l'Universite de Quebec, Montreal, pp 129–153

Müller DK, Jansson B (2007) The difficult business of making pleasure peripheries prosperous: perspectives on space, place and environment. In: Müller DK, Jansson B (eds) Tourism in peripheries: perspectives from the far north and south. CABI, Wallingford, pp 3–18

Naturvårdsverket (2010) Miljömål i fjällandskapet: En syntes av problemställningar knutna till förvaltningen av en begränsad resurs. Rapport 6366. Stockholm

Puhakka R, Sarkki S, Cottrell SP, Siikamaäki P (2009) Local discourse and international initiatives: sociocultural sustainability of tourism in Oulanka National Park, Finland. J Sustain Tour 17(5):529–549

Sandell K (1995) Access to the 'North' – but to what and for whom? Public access in the Swedish countryside and the case of a proposed national park in the Kiruna mountains. In: Hall CM, Johnston ME (eds) Polar tourism: tourism in the Arctic and Antarctic regions. Wiley, Chichester, pp 131–146

Sandell K (2007) Access, tourism and democracy: a conceptual framework and non-establishment of a proposed national park in Sweden. In: Müller DK, Jansson B (eds) Tourism in peripheries: perspectives from the far north and south. CABI, Wallingford, pp 141–153

Sörlin S (1988) Framtidslandet: Debatten om Norrland och naturresurserna under det industriella genombrottet. Carlsson, Stockholm

Stewart EJ, Draper D (2006) Sustainable cruise tourism in Arctic Canada: an integrated coastal management approach. Tour Mar Environ 3(2):77–88

Timothy D (2010) Contested places and the legitimization of sovereignty through tourism in polar regions. In: Hall CM, Saarinen J (eds) Tourism and change in polar regions: climate, environment and experience. Routledge, London, pp 288–300

Wall Reinius S (2009) A ticket to national parks? Tourism, railways, and the establishment of national parks in Sweden. In: Frost W, Hall CM (eds) Tourism and national parks: international perspectives on development, histories and change. Routledge, London, pp 184–196

Wall Reinius S, Fredman P (2007) Protected areas as attractions. Ann Tour Res 34(4):839–854

Zackrisson A, Sandell K, Fredman P, Eckerberg K (2006) Tourism and protected areas: motives, actors and processes. Int J Biodivers Sci Manag 2(4):350–358

Newspaper Articles

Dagens Nyheter (2008-05-05) Motståndet växer mot nationalpark i Vindelfjällen (Annika Rydman)

Dagens Nyheter (2010-04-26) Allas vår änglamark (NN)

Dagens Samhälle (2007-10-25) Släpp in oss i nationalparksarbetet! (Bengt-Urban Fransson)
Dagens Samhälle (2008-03-27) Kluven inställning till naturskydd (Lars Hugo)
Göteborgs Posten (2011-01-28) Utförsåkningens svenska Mekka (NN)
Hallands Nyheter (2007-09-07) Blackstone Nickel vill bryta i Vindelfjällen (NN)
VF (2008-02-13) Frågan om nationalpark delar Storumanborna (Elin Olsson)
VF (2008-02-14) Storuman säger nja till nationalpark (Elin Olsson)
VF (2008-02-19) Samebyar protesterar (Erik Isberg)
VF (2008-06-24) Beslut om förstudie kring nationalpark dröjer (Jessica Tjällman)
VF (2008-12-03) Vindelfjällen med i nationalparksplanen (Tor Larsson)
VK (2007-04-27) Vindelfjällen föreslås bli nationalpark (Bertholof Brännström)
VK (2007-04-28) Privat mark i farozonen (Allan Fjällström)
VK (2007-06-26) Öjeryd ordförande i renhållningsnämnden efter knapp valseger (Bertholof Brännström)
VK (2007-08-15) Motstånd mot ny nationalpark (NN)
VK (2007-08-23) Vindelfjällens nationalpark inte störst (Allan Fjällström)
VK (2007-09-11) Nationalparksplaner engagerar sorselebor (NN)
VK (2007-09-12) Vindelfjällen som ny nationalpark? (NN)
VK (2007-09-13) Storuman säger nej till förstudies om nationalpark (Roland Tidström)
VK (2007-10-18) Länsstyrelsen kommer att säga nej (Bertholof Brännström)
VK (2007-12-07) Stora fynd av malm i området (Roland Tidström)
VK (2007-12-28) Ja till nationalparker, men sametinget vill ha inflytande (Bertholof Brännström)
VK (2008-02-20) Kräver stopp för provborrning (Bertholof Brännström)
VK (2008-03-13) Nationalpark i Vindelfjällen katastrof for de bofasta ("Fjällbo")
VK (2008-04-21) Nej till nationalpark i Vindelfjällen (Sven Hermansson)
VK (2008-04-28) Lovar lyssnar på kommunen (Roland Tidström)
VK (2008-04-28) Nybildat motstånd mot nationalpark (Roland Tidström)
VK (2008-05-17) Sorsele har ingen vetorätt (Sven Björkland)
VK (2008-05-31) Vi håller vad vi har lovat: Länsstyrelsen lovar lyssnar på majoriteten om nationalpark i Vindelfjällen (Bertholof Brännström)
VK (2008-06-25) Beslutsvånda i Sorsele (NN)
VK (2008-09-11) Skepsis mot garantier (Bertholof Brännström)
VK (2008-09-23) Förstudie kan ge svar på ortsbornas frågor (Bertholof Brännström)
VK (2008-09-23) Nationalpark hindrar bygdens utveckling (Bertholof Brännström)
VK (2008-10-04) "Vi ger inte upp" (Sven Björkland)
VK (2009-12-04) "Beslutet var väntat" (Sven Björkland)
VK (2009-01-15) "Vi har inte sagt ja till nationalpark (Roland Tidström)
VK (2009-01-17) KD motståndare till nationalpark (Roland Tidström)
VK (2009-09-11) Nationalpark i Vindelfjällen: frågan om vägen Sorsele-Ammarnäs bör lyftas till regional/nationell nivå (Tomas Staafjord)

Chapter 14
Indigenous People: Discussing the Forgotten Dimension of Dark Tourism and Battlefield Tourism

Raynald H. Lemelin, Anna Thompson-Carr, Margaret Johnston, Emma Stewart, and Jackie Dawson

Abstract Dark tourism and battlefield tourism have tended to focus on the social aspects of visits, such as marketing and interpretation, yet largely ignore the cultural and spiritual dimensions of these sites of death and loss. This chapter addresses these shortcomings through an examination of two well-known indigenous battlefields sites and outlines the potential lessons learned from these sites for the Canadian north. The goals of this chapter are both to (a) provide a contextual overview of issues related to the recognition and development of indigenous sites of death, loss, and battlefields, including cultural dissonance, and (b) explore these processes with reference to the Canadian North. We will address some of the limitations that have been noted in dark and battlefield tourism, including the tendency to focus on European sites or sites of colonial "victories" which perpetuate grand "colonial" narratives, while arguing for the need to recognize commemorative, cultural, and spiritual aspects. We also provide a rationale as to why most sites in Northern Canada have been excluded from the dark and/or battlefield narratives.

R.H. Lemelin, Ph.D. (✉) • M. Johnston, Ph.D.
School of Outdoor Recreation, Parks and Tourism, Lakehead University, Thunder Bay, ON P7B 5E1, Canada
e-mail: rhlemeli@lakeheadu.ca; mejohnst@lakeheadu.ca

A. Thompson-Carr, Ph.D.
Department of Tourism, School of Business, University of Otago, Dunedin, New Zealand
e-mail: anna.thompson@otago.ac.nz

E. Stewart, Ph.D.
Faculty of Environment, Society & Design, Lincoln University, PO Box 84, Lincoln 7647, Canterbury, New Zealand
e-mail: emma.stewart@Lincoln.ac.nz

J. Dawson, Ph.D.
Department of Geography, University of Ottawa, Ottawa, ON K1N 6N5, Canada
e-mail: jackie.dawson@uottawa.ca

D.K. Müller et al. (eds.), *New Issues in Polar Tourism: Communities, Environments, Politics*, DOI 10.1007/978-94-007-5884-1_14,

Keywords Battlefield tourism • Colonialism • Dark tourism • Indigenous peoples • New Zealand • Northern Canada • Warfare tourism

14.1 Introduction

Well-known dark tourism and battlefield sites around the world include Gettysburg and the Battle of the Little Bighorn in the United States; the Plains of Abraham and Grosse Île and the Irish Memorial in Canada; the White Tower in England; Culloden and Bannockburn, Scotland; the Cu Chi Tunnels in Vietnam; and Lucknow Residency in India. The national and international interest in preserving such historic sites normally requires management agencies to include and even promote tourist visitation to these sites of death and loss, resulting in the development of tourist attractions featuring marketing, interpretation, site access, and related visitor services. In Northern Canada such site recognition and development has largely been reserved for Euro-Canadian sites, or those involving European exploration, for example, those sites related to the quest for the Northwest Passage in Northern Canada.

Though management agencies have been willing to promote death and tragic events at some sites, they have often been less willing to discuss the colonial and exclusionary aspects of these sites (Hannam 2006; Schwenkel 2006). For example, in North America, various strategies have been used to remove, expropriate, and deny the access of various groups at national parks in the United States and Canada: women (Kaufman 2006), indigenous peoples (Denzin 2008), African Americans (Erickson et al. 2009), African Canadians (Killam 1993), Jewish people (Killam 1993), and French Canadians (Bernier 2009). During World War II, well-known national parks in Western Canada (Riding Mountain NP in Manitoba and Jasper NP in Alberta) were also used to incarcerate Japanese-Canadians prisoners of war and conscientious objectors such as Mennonites, Seventh Day Adventists, and Hutterites. These colonial and exclusionary practices are no longer promoted in North America, yet with some exceptions (see Lemelin and Baikie 2011; Pullar 2008), the legacy of this cultural imperialism continues, for example, in the absence of these voices in parks management or museum interpretation strategies.

These current practices also indirectly reinforce the grand narratives of colonialism, where hegemonic ideologies of progress, democracy, "individualism, and moral goodness [...] remain unchallenged" (Schwenkel 2006, p. 6). In many instances, such approaches have resulted in historical and dialectical narratives that are selective, partial, biased, and distorted (Henderson 2000; Nasson 2000; Tunbridge and Ashworth 1996) and have contributed (either directly or indirectly) to cultural dissonance (Hannam 2006; Winter 2011). For example, in the "first Indian rebellion" of India, "violence perpetrated by British soldiers was seen as normal, rational, and even natural, whereas violent acts perpetrated by Indian soldiers were viewed as extreme, irrational, and even unnatural" (Hannam 2006, p. 204). In instances where colonial forces were defeated in North America, such as the Battle of the Little Bighorn, the Battle of Sitka in Alaska, or Batoche in Canada, the "stand" or tactics by the colonial armies were defined as heroic and defiant, whereas the indigenous forces were described as using "shifty" and "underhanded tactics." In other instances, "skirmishes" between colonial forces and indigenous peoples were deemed of little consequence, when compared to the large engagements on the European and North American continents (i.e., Battle of Waterloo, Gettysburg, Battle of 1812, WWI, and WWII) (Prideaux 2007). Furthermore, rarely discussed in the literature are sites of massacres of indigenous people like the massacre of the Suspiat on Kodiak Island in 1784 (Black 1992, 2004; Pullar 2008).

This chapter begins with a general discussion on indigenous battle sites in dark and warfare tourism. Following, we then examine the Battle of Little Bighorn in the USA and the battle of Batoche in Canada in order to set the scene for a discussion of battlefields for those sites of loss in the Northern Canada, and possible management options to that sites from New Zealand are provided. In light of these examples, we focus on reexamining the concept of dark tourism, providing a broader understanding of battlefield tourism, and outlining research needs. The goals of this chapter are to (a) provide a contextual overview of issues related to the recognition and development of indigenous sites of death, loss, and battlefields, including cultural dissonance, and (b) explore these processes with reference to Northern Canada and other polar countries.

14.2 Coming to Terms with Terms

Since the 1990s, terms such as atrocity heritage (Ashworth 2002), battlefield tourism, black spot tourism (Rojek 1993), dark tourism (Lennon and Foley 2000; Sharpley and Stone 2009), gray tourism (Sharpley 2005), le tourisme macabre (Hernandez 2008), morbid tourism (Blom 2007), thanatourism (Seaton 1996; Seaton and Lennon 2004), tragic tourism (Lippard 1999), and warfare tourism (Gordon 1998) have been used to make sense of the tourism consumption of human death, loss, and disaster. As this study examines issues related to indigenous battlefields, the terms warfare and battlefield tourism will be used here. Unlike the traveler who seek peace and tranquility, according to Schwenkel (2006, p. 4), the warfare tourist "is driven by the desire to see, experience, and understand mass destruction and violence."

14.2.1 Warfare Tourism and Indigenous Sites

Sites associated with warfare tourism include war memorials, cemeteries, burial sites of military leaders, war museums, warships and aerial display of WWII bombers, peace parks, battle reenactments, and battlefield tours (Dunkley et al. 2011). Motivated by factors such as celebrations, reflections, and reinterpretations of the conflicts through memoirs, novels, movies, comic books, computer games, music (Offenstadt 2010; Schwenkel 2006), and autohistory told from indigenous perspectives (Sioui 1992). There are a variety of reasons why tourists visit these types of sites including commemoration and pilgrimage, entertainment, and education (Dunkley et al. 2011; Lloyd 1998; Offenstadt 2010). Research indicates that battlefield tourism experiences can be emotionally cathartic, spiritual, meaningful, transformative, and restorative (Dunkley et al. 2011; Lloyd 1998).

Visitors to these sites originate in "many cultural and ethnic groupings however and as such, these sites can also [become] sites of contested remembrance which disrupt the nature of individual collective and national memory" (Dunkley et al. 2011, p. 866). There are numerous examples of battlefields and contested areas with indigenous significance throughout the world, where colonial powers used exploration, expropriation, and killing to achieve extensions of their empires. Some of these sites are memorialized, but many are not. For example

> Skirmishes between European settlers and Aboriginals, defending their tribal lands are largely forgotten and the sites rarely visited or even acknowledged. Not surprisingly the level of recognition given to the Australian battlefield sites compared to overseas battlefield field sites where Australians died is small. Thus, recognition of the nation's military history occurs primarily through museums, memorials and monuments rather than though battlefield visitation. (Prideaux 2007, p. 23)

In some cases, sites of "rebellion" become national monuments (e.g., Lucknow in India) or the symbolic "heart" of a nation (e.g., Batoche for the Métis in Canada). In other instances, little known narratives are heard once again, sometimes for the first time in many generations. An exception to this "silence" is the management of battlefields in Aotearoa/New Zealand. Though such sites are also contested, the existence of the Treaty of Waitangi and the close relationships Māori have maintained with such sites and, despite colonization, have resulted in many sites of battle gaining national recognition (Prickett 2002).

Further, as Hannam (2006) explains, little published research has examined the contested meaning of warfare tourism in colonial and postcolonial contexts. Overall, warfare tourism "in non-western cultures remains a much neglected area of research, worthy of further research" (Hannam 2006, p. 211).

In order to explore the dialectical issues surrounding recognition, interpretation, and cultural dissonance, we now discuss two battlefields. Due to our familiarity with North America, we have selected sites in Canada and in the United States, both of which are well known, commemorated, and documented, thereby providing ideal cases. Our interest is in understanding these processes for reference to Arctic Canada, where land claims and comanagement strategies adopted by indigenous groups have in some case forced management agencies to address the issue of cultural dissonance; in other instances, these issues remain peripheral and have yet to be addressed.

14.2.2 The Battle of Little Bighorn, 1876, Montana

On June 25th, 1876, Lt. Col. George Armstrong Custer and the 7th Cavalry, aided by Arikara and Crow scouts, attacked a large village located along the Little Bighorn

River in what would later become the state of Montana, which was being defended by Sioux and Cheyenne warriors and led by Sitting Bull and Crazy Horse. “By the end of the day, the Sioux and Cheyenne had defeated the 7th Cavalry, and all of the men under Custer’s immediate command lay dead (approximately 263 soldiers and attached personnel)” (Buchholtz 1998, p. 113). The site first protected as a US National Cemetery in 1879 was later designated as the Custer Battlefield National Monument. After considerable debate and in order to recognize all of the actors involved in this battle, the Custer Hill National Monument was renamed the Little Bighorn Battlefield National Monument in 1991. Each year, 300,000 tourists visit the battlefield grounds which consist of a visitor center, a mass cavalry grave, additional cemeteries, monument, and seasonal guided tours by parks staff, and tour packages are offered through the Wind in Feather tours given by the Crow Nation. Battle reenactments continue to attract thousands of visitors each year and generate millions of dollars (Buchholtz 1998, 2005).

Though controversial at the time, the site’s renaming better reflects the actual events that occurred during the battle (Buchholtz 1998). Along with the renaming, a memorial dedicated to Sioux and Cheyenne participants was also erected on the ridge near the 7th Cavalry monument (Buchholtz 1998). The visitor center also provides educational materials about efforts to repatriate Cheyenne skeletal remains from the Smithsonian and Peabody museums (Buchholtz 1998). This case is significant in demonstrating the belated recognition of the roles played by Native Americans and indicating efforts made by the management agency to provide a fuller history of the battle and cultural issues of relevance for the people involved.

14.2.3 The Northwest Resistance of 1885 at Batoche, Saskatchewan

Encroaching settlements by nonindigenous settlers and development projects financed by the Canadian government led to an armed Métis uprising in 1885 (Boyden 2010). After some initial success against federal troops, the outnumbered Métis were besieged at Batoche in what would later become the province of Saskatchewan and subsequently defeated (Boyden 2010; Combet and Toussaint 2009; Hutton 1996). Their leader, Louis Riel, who had led a previous uprising in what would later become the province of Manitoba, was captured, convicted of treason, and hanged in Regina (Boyden 2010). Gabriel Dumont, the Métis general, received amnesty and was permitted to return to Batoche (Combet and Toussaint 2009; Hutton 1996). Though the village was abandoned around 1900, a national historic site was established in 1923 and is managed by Parks Canada. This site contains a visitor information center, the remains of the St. Antoine de Padouche church and its rectory, shallow riffle depressions, the Métis mass grave, and the tomb of Gabriel Dumont (Ryan 2007). The designation of the Batoche National Historic Site has created tensions between the federal agency mandated to manage the site and the Métis Nation, especially since the Métis claim ownership of Batoche as their ancestral capital, and have gathered there each summer since the 1970s in the tens of thousands to celebrate Back to Batoche Days (Hutton 1996). This example illustrates a source of conflict that can arise when management concerns are different from those of indigenous peoples. This discussion will be reexamined later.

14.2.4 Conflicts in Greenland, Alaska, and Canada

Conflicts between Viking settlers and Greenland Eskimo were documented around the twelfth century AD (Diamond 2011). Various conflicts were noted between Russian explorers and indigenous people in Alaska in the late eighteenth century. Documented conflicts include the massacre of the Suspiat on Kodiak Island in 1784 (Black 1992, 2004; Pullar 2008), the Battle of Sitka in the early nineteenth century (Dauenhauer et al. 2008), and the Nulato massacre in 1851 (Reedy-Maschner and Maschner 1999).

Despite various conflicts, mutinies (e.g., Henry Hudson in 1610), tragedies, and cultural misunderstandings (Mancall 2010), most historic sites of tragic events and death in the Northern Canada have been dedicated to European explorers. These sites include the various “discoveries” and deaths associated with the exploration of and quest for the Northwest Passage throughout the eighteenth and nineteenth centuries (e.g., Beechey Island, O’Reilly Island and the two ships, the Terror and Erebus from Franklin’s Expedition). One national historic site, established in 1978, and also a territorial park, does showcase a conflict between Dene and Inuit people at the Kugluk/Bloody Falls. During this particular event, as the Dene were escorting Samuel Hearn across in the Northern Canada in 1771, they came upon a small aggregation of Inuit families at the falls; the Dene fell upon the Inuit and massacred over 20 members of the party (Brand 1992; McGrath 1992). In the late 1950s, the famines at Garry Lake (Hanningajuq) NWT took the lives of numerous indigenous inhabitants and led to the relocation of the remaining survivors at Baker Lake (Tester and Kulchyski 1994). Apart from the crosses, very little discussion or interpretation of this event and this site is provided.

Also largely overlooked throughout most of the north is how the threat of war during the cold war era also justified the forced relocation and sedentarization of various indigenous groups (i.e., Dene, Cree, Inuit) throughout the Canadian sub-Arctic and Arctic (Marcus 1995; Tester and Kulchyski 1994). Scholars, both Euro-Canadian and aboriginal, have critiqued the dominant colonial discourse presented in these interpretation strategies and suggested that such approaches are changing because indigenous people are being provided with greater control over the management of these areas (through land claims and the devolution of governments) (see Lemelin and Baikie 2011).

14.3 Discussion and Conclusion

The similarities between sites commemorating the battles of the Little Bighorn and Batoche are striking. Both involve indigenous victories over colonizing forces, an official commemoration of the battlefield, and interpretation that emphasized one side of the story. Indeed, it could be argued that until the mid-twentieth century, these sites, through existing interpretation strategies, reinforced the grand narrative of colonialism, thereby contributing to cultural dissonance. However, ongoing defiance and the presentation of alternative interpretations in the case of the Battle of the Little Bighorn resulted in a name change, the recognition of Sioux and Cheyenne warriors, and discussions of key issues regarding the repatriation of skeletal remains. The ongoing success of the "Back to Batoche Days" (see Back to Batoche (n.d.) website http://www.backtobatoche.org/) demonstrates that the Métis Nation is alive and well and will continue to remind federal and provincial agencies that this national historic site was a battle fought in the name of freedom against an imperialistic force, and will remain the spiritual capital of the Métis Nation.

A number of challenges associated with the management and interpretation of these two battlefields persist. One is the inadequate attention paid to the military prowess of three war leaders: Sitting Bull, Crazy Horse, and Gabriel Dumont, who by sheer audacity and skills defeated large colonial armies (Boyden 2010; Matthiessen 1992).

As well, both sites currently promote sanitized and simplistic models of social relations, implying that group identities and battle commitments marked by the indigenous/European dichotomy are static and inevitable (Buchholtz 1998). Such interpretations fail to recognize that some indigenous groups like the Arikara and Crow in the USA actively supported the 7th Cavalry at the Battle of Little Bighorn and some First Nations in Canada did not support the Métis resistance at Batoche. Furthermore, this interpretation fails to recognize that many French Canadians throughout Canada supported Métis resistance and also see Batoche as a sacred site.

Some researchers have suggested that the lack of commemoration of a battlefield is indicative of colonial hegemony (see, e.g., Ryan 2007). In the case of the Battle of the Little Bighorn, the recognition of the departed, place renaming, apologies, tourism, and management strategies have done much to address past wrongdoings. For Batoche, tensions and interpretive differences between federal agencies and indigenous people remain, while in Northern Canada, few sites recognize or even discuss conflicts between indigenous peoples or between indigenous people and explorers. Indeed, much of the narrative is dedicated to Europeans explorers and Euro-Canadians. However, this aboriginal erasure from the cultural narrative is being challenged through land claims, culturally sensitive interpretive strategies in museums, and the comanagement of parks and historic sites at the territorial and national levels (Lemelin and Baikie 2011; Pullar 2008). Therefore, Ryan's (2007) conclusion that all indigenous sites should be recognized and commemorated is insufficient, for commemoration of battlefields and sites of death and loss in some case can require the establishment of monuments, whereas in some places, such as New Zealand, markers are sometimes sufficient. In others, recognition of the field where combat occurred is needed. These differences suggest that signage and management strategies (even with the best intentions) can, without proper consideration of contextual cues and appropriate consultation (i.e., "consultation with tangata whenua" in New Zealand), show disrespect to and dishonor local customs and may actually end up doing more harm than good. The solution is to permit 'polyphonic representations allow the heritage of these sites to be interpreted in a variety of ways' (Hannam 2006). As Offenstadt (2010) and Lloyd (1998) explain, the act of visiting these sites (even if they are not commemorated) is a dialectical process that both mourns and honors, that acknowledges both the past and the present, and that recognizes the victims and the healers. Similar strategies to those employed in New Zealand, a Commonwealth country with a history of colonization of the Māori (tangata whenua – the people of the land), may also provide culturally sensitive and inclusive approaches to managing these issues.

In New Zealand, the Department of Conservation (DoC) manages most known sites of conflict, many of which remain in relatively undeveloped or rural areas. Historic places such as Ruapekapeka Pa in the North Island where the last conflicts between colonizing British troops and local Māori of the Ngapuhi iwi (tribe) occurred in the mid-1840s are now sites of visitation (Prickett 2002; Department of Conservation 2003). The management of these sites requires consideration of specific cultural values (tikanga Māori) such as the tapu (forbidden or sacred nature) or wairua (spirit) of battle sites where ancestors of today's tangata whenua passed away (ibid.). In New Zealand, such sites may also be collaboratively managed by the Department of Conservation (DoC), New Zealand Historic Places Trust (NZHPT), and local iwi (Māori tribes living by or associated with the area). When sites are under DoC management, extensive consultation occurs with the relevant iwi representatives usually through local runanga (tribal administrative offices) and with elders, who are often Māori leaders. In all of these cases, the act of visiting or discussing these sites is a dialectical process of consultation and shared management processes (Lloyd 1998).

New Zealand management of indigenous sites of battle is increasingly a process of collaborative practices with management plans being administered by the department or specific management trusts. An example of this is the Ruapekapeka Pa Management Trust which consists of representatives who are the descendants of the Māori iwi members who built and fought at the site. The Trust works alongside DoC "To ensure that Ruapekapeka Pa be maintained, protected, and preserved for the future and that Ruapekapeka stories are made available to the public ensuring that the taha wairua of the site is retained for the benefit and education of all New Zealanders" (Department of Conservation 2011). Thus, the interpretation trails available at such sites become a focus of visitation for learning experiences by domestic and international visitors. Elsewhere in New Zealand's North Island, Opepe Historic Reserve and Te Porere Redoubt, near Lake Taupo, are two other nationally significant sites of conflicts or battle where both Māori and non-Māori died. At both sites, visitation is actively managed through hardened trails and interpretation, and management plans and archaeological reports inform the DoC about the site's preservation and appropriate visitor management (though impact such as visitor-induced trampling and erosion still occur) (Williams 1999; Ritchie and Barr n.d.). DoC and NZHPT normally ensure consultation and representation of local Māori iwi in decision-making surrounding such sites but such management practices can still be contentious with modern Māori often regarding any government "ownership" of, or legislative responsibility for, these historic lands as perpetuating colonialism in the twenty-first century (despite best practice consultation and comanagement practices occurring).

Canada could manage battle sites for tourism or visitation purposes by adopting practices similar to those in New Zealand, especially collaborative or comanagement practices, but this will require much investment of time and resources (and willpower) to ensure sincere and effective relationships are built between relevant government agencies and indigenous groups. Canada has the added challenges of, firstly, the tyranny of distance across the vast Canadian North and secondly many distinctive indigenous cultural groups (whereas New Zealand Māori, – while consisting of numerous tribes – share common cultural values, genealogy, history, and language).

This study provided a case study of two battlefields in North America and the aboriginal erasure of indigenous people in historic sites in Northern Canada and New Zealand. However, such erasure of indigenous voice can be addressed through meaningful representation as has been the case in New Zealand. It is hope that this chapter is another small but worthy contribution to this field of research and illustrated the contextual nuances (both historical and contemporary) at play. While battles and conflicts between indigenous peoples and colonizers and settlers in Greenland and Alaska were mentioned, they were not addressed in any great detail. Also overlooked is the commemoration of these sites in the north and how they differ from those elsewhere. Further research on these sites and associated commemorative strategies is required. Thus, we encourage indigenous and nonindigenous authors from these and other nations to engage in these discussions, and perhaps with consensus, and address the issue of cultural dissonance. For as Bowman and Pezzulo (2010) argued, by recognizing and discussing past wrongdoings, these sites can be transformative and restorative, challenging and engaging, and healing and empowering. It is, therefore, not just about preserving history but also about shaping the present and the future.

References

Ashworth GJ (2002) Holocaust tourism: the experience of Kraków-Kazimierz. Int Res Geogr Environ Educ 11(4):363–367

Back to Batoche Days (n.d.) http://www.backtobatoche.org/. Accessed 25 Aug 2011

Bernier L (2009) La bataille de Forillon. Fides, Saint-Laurent

Biran A, Poria Y, Oren G (2011) Sought experiences at dark heritage sites. Ann Tour Res 38(3):820–841

Black LT (1992) The Russian conquest of Kodiak. Anthropol Pap Univ Alsk 24(1–2):165–182

Black LT (2004) Warriors of Kodiak: military traditions of Kodiak islanders. Arct Anthropol 41(2):140–152
Blom T (2007) Morbid tourism: the case of Diana, Princess of Wales and the Althorp House. In: Long P, Palmer NJ (eds) Royal tourism: excursions around monarchy. Channel View, Clevedon, pp 142–158
Bowman MS, Pezzulo PC (2010) What's so 'dark' about 'dark tourism'?: Death, tours, and performance. Tour Stud 9(3):187–202
Boyden J (2010) Louis Riel & Gabriel Dumont. Penguin Canada, Toronto
Brand MJ (1992) Samuel Hearne and the massacre at Bloody Falls. Polar Rec 28:229–232
Buchholtz D (1998) The battle of the Little Bighorn: History, identity, and tourism in the 1990s. In: Lew AA, van Otten GA (eds) Tourism and gaming on American Indian lands. Cognizant Communication Corporation, New York, pp 113–127
Buchholtz D (2005) Cultural politics or critical public history? Battling on the Little Bighorn. J Tour Cult Chang 3(1):18–35
Combet D, Toussaint I (2009) Gabriel Dumont: Souvenirs de resistance d'un immortel de l'Ouest l'Editions Cornac, Montreal, QC
Dauenhauer NM, Dauenhauer R, Black LT (eds) (2008) Anooshi Lingit Aani Ka, Russians in Tlingit America: the Battle of Sitka, 1802 and 1804. University of Washington Press, Seattle
Denzin N (2008) Searching for Yellowstone: race, gender, family, and memory in the postmodern west. Left Coast Press, Walnut Creek
Department of Conservation (2003) Ruapekapeka Pa: a site guide. Department of Conservation, Northland
Department of Conservation (2011). Ruapekapeka Pa Management Trust. At http://www.doc.govt.nz/getting-involved/volunteer-join-or-start-a-project/join-a-group/northland/ruapekapeka-pa-management-trust/. Accessed 10 June 2011
Diamond J (2011) Collapse: how societies choose to fail or succeed, Rev edn. Penguin, Toronto
Dunkley R, Morgan N, Westwood S (2011) Visiting the trenches: exploring meanings and motivations in battlefield tourism. Tour Manag 32:860–868
Erickson B, Johnson CW, Kivel RD (2009) Rocky Mountain National Park: history and culture as factors in African American park visitation. J Leis 41(4):529–545
Gordon BM (1998) Warfare and tourism: Paris in World War II. Ann Tour Res 25(3):616–638
Hannam K (2006) Contested representations of war and heritage at the Residency, Lucknow, India. Int J Tour Res 8(3):199–212
Henderson JC (2000) War as a tourist attraction: the case of Vietnam. Int J Tour Res 2:269–280
Hernandez J (2008) Le tourisme macabre la Nouvelle-Orléans après Katrina: Résilience et mémorialisation des espaces affectés par des catastrophes majeures. Norois 208:61–73
Hutton DA (1996) Back to Batoche: A Cultural centre for the Métis Nation of Saskatchewan. Master's thesis. University of British Columbia, Vancouver, BC
Kaufman PW (2006) National parks and the woman's voice. University of New Mexico Press, Albuquerque
Killam G (1993) Protected places: a history of Ontario's provincial parks system. Dundurn Press Limited, Toronto
Lemelin RH, Baikie G (2011) Our home and native land: recognizing the socio-cultural dimensions of last chance tourism. In: Lemelin RH, Dawson J, Stewart E (eds) Last chance tourism. Routledge, New York
Lennon J, Foley M (2000) Dark tourism: the attraction of death and disaster. Continuum, London
Lippard L (1999) Off the beaten track: tourism, art, and place. The New Press, New York
Lloyd D (1998) Battlefield tourism: pilgrimage and the commemoration of the Great Ware in Britain, Australia and Canada, 1919–1939. Berg, New York
Mancall PC (2010) Fatal journey: the final expedition of Henry Hudson. Basic Books, New York
Marcus AR (1995) Relocating Eden: the image and politics of Inuit exile in the northern Canada. Dartmouth College, Hanover
Matthiessen P (1992) In the spirit of crazy horse. Penguin, Toronto

McGrath R (1992) Samuel Hearne and the Inuit oral tradition. Stud Can Lit/Études en littérature canadienne 18(2). On-line document available at: http://journals.hil.unb.ca/index.php/SCL/article/viewArticle/8186/9243. Retrieved 5 June 2011
Nasson B (2000) Commemorating the Anglo-Boer in post-apartheid Africa. Radic Hist Rev 78:149–165
Offenstadt N (2010) 14–18 Aujourd'hui: La grande guerre dans la France contemporaine. Odile Jacob, Paris
Prickett N (2002) Landscapes of conflict. Random House, Auckland
Prideaux B (2007) Echoes of war: battlefield tourism. In: Ryan C (ed) Battlefield tourism: history, place and interpretation. Elsevier, Amsterdam, pp 17–27
Pullar GL (2008) Repatriation, cultural revitalization, and Indigenous healing in Alaska. In: Gabriel M, Dahl J (eds) UTIMUT: past heritage – future partnerships: discussions on repatriation in the 21st century. IWGIA, Copenhagen, pp 108–115
Reedy-Maschner KL, Maschner HDG (1999) Marauding middlemen: western expansion and violent conflict in the subarctic. Ethnohistory 46(4):703–743
Ritchie NA, Barr C (n.d.) Te Porere historic reserve management plan. Department of Conservation, Hamilton
Rojek C (1993) Ways of escape: modern transformations of leisure and travel. Macmillan, London
Ryan C (2007) The battles of Rangiriri and Batoche: amnesia and memory. In: Ryan C (ed) Battlefield tourism: history, place and interpretation. Elsevier, Amsterdam, pp 87–97
Schwenkel C (2006) Recombinant history: transnational practices of memory and knowledge production in contemporary Vietnam. Cult Anthropol 21(1):3–30
Seaton AV (1996) Guided by the dark: from thanatopsis to thanatourism. Int J Herit Stud 2: 234–244
Seaton AV, Lennon J (2004) Moral panics, ulterior motives and alterior desires: thanatourism in the early 21st century. In: Singh TV (ed) New horizons in tourism: strange experiences and stranger practices. CABI, Wallingford, pp 63–82
Sharpley R (2005) Travels to the edge of darkness: towards a typology of dark tourism. In: Ryan C, Page S, Aitken M (eds) Taking tourism to the limits: issues, concepts and managerial perspectives. Elsevier, Oxford, pp 217–228
Sharpley R, Stone PR (eds) (2009) The darker side of travel: the theory and practice of dark tourism. Channel View, Bristol
Sioui G (1992) For an Amerindian autohistory. McGill University Press, Montreal
Strange C, Kempa M (2003) Shades of dark tourism: Alcatraz and Robben Island. Ann Tour Res 30(2):386–405
Tester F, Kulchyski P (1994) Tammarniit (mistakes): Inuit relocation in the eastern Arctic, 1939–63. UBC Press, Vancouver
Tunbridge JE, Ashworth GJ (1996) Dissonant heritage: the management of the past as a resource in conflict. Wiley, New York
Williams A (1999) Threats to archaeological features within the redoubt and village area, Opepe Bush Scenic and Historic Reserve, Taupo. Science and Research internal Report No. 168. Department of Conservation, Wellington
Wilson JZ (2008) Prison: cultural memory and dark tourism. Peter Lang Publishing, New York
Winter C (2011) Battlefield tourism and Australian nationalism. In: Frew E, White L (eds) Tourism and national identities: an international perspective. Routledge, London/New York, pp 176–189

Chapter 15
New Issues in Polar Tourism: Conclusion

Linda Lundmark, Raynald H. Lemelin, and Dieter K. Müller

Abstract Although the goal of this book was not to pursue a distinct research question but rather illustrate the multitude of thematic issues that are currently being investigated in the polar Norths, two topics, the governance of ecological resources and the ways in which polar communities manage to create agency through various development strategies, emerged throughout the book. In order to illustrate this agency, the content of this book has been divided into three parts: conceptualizing polar tourism and polar regions, politics and the environment, and business and community perspectives, thus without a separation between Arctic and Antarctic research. In this concluding chapter, a summary of the issues in polar tourism highlighted in this book is made, and a comment on the current state of the research field is offered, with some suggestions for future research.

Keywords Agency • Conceptualizing the polar • Destination development • Governance • Tourism impacts

15.1 What is so Special About the Polar?

Separating the polar research arena from other areas of research is not easy, especially when we consider the various disciplines involved in tourism research in and on the polar norths or the various definitions used to define this area. Yet as this volume illustrates, disciplinary, interdisciplinary, and transdisciplinary approaches

L. Lundmark, Ph.D. • D.K. Müller, Ph.D. (✉)
Department of Geography and Economic History, Umeå University, SE-90781 Umeå, Sweden
e-mail: linda.lundmark@geography.umu.se; dieter.muller@geography.umu.se

R.H. Lemelin, Ph.D.
School of Outdoor Recreation, Parks and Tourism, Lakehead University, 955 Oliver Road, Thunder Bay, ON P7B 5E1, Canada
e-mail: rhlemeli@lakeheadu.ca

D.K. Müller et al. (eds.), *New Issues in Polar Tourism: Communities, Environments, Politics*, DOI 10.1007/978-94-007-5884-1_15,

to polar tourism are promoting cooperative and collaborative research approaches and providing new avenues for problem-solving. Although the boundaries set for the polar are largely dependent on what people perceive as being polar, an additional factor to be considered might be the communities themselves and what they perceive as being polar areas.

For example, Brouder and Lundmark (Chap. 10, this volume) stress the special characteristics of the Arctic for tourism development and adaptation to climate change in particular, highlighting the geographical characteristics and relationships that exist within and between polar areas and other parts of the world. Furthermore, science plays a central role in the shaping of the social, cultural, political, and geographical understandings of polar areas. Scientists are thus involved in the process of creating and recreating territorial, geographical, and even cultural boundaries. Such constructions, explains Müller (Chap. 13, this volume), have local and regional implications for communities and associations located in the polar norths. Because of this, consideration regarding the legal and ethical implications on local people from such definitions, often generated through research, is required.

In their competition for tourism, destinations, it has been argued, will often seek to brand or market the uniqueness of their locals or resources, often at the expense of regional economic strategies. Although not the specific focus of the book, some chapters discuss the implications of such competitive strategies and the benefits (or lack thereof) of the trickle-down effect on local economic development. In tourism marketing strategies, the construction of place or such locations as sub-Antartica suggest Shelton (Chap. 4, this volume), this approach is an important step in order to deconstruct stereotypes and appropriately market a destination. The implications from such approaches are that local communities can, in some cases, influence which images should be "broadcasted" and determine hence how the product is marketed. The type of tourists attracted to these images, and the often unexpected outcomes from these visitor expectations, is discussed by Mäkitie and Ylisirniö (Chap. 11, this volume).

15.2 Governing Polar Areas

From assessments to monitoring schemes, from legal mechanisms to intra-jurisdictional policies, there are many ways in which governments and various political agencies govern polar areas. In some cases, local and regional governments are included in these approaches, while in others they are excluded from. Local and regional strategies for development and cooperation often involve the so-called Triple Helix Model, in which collaboration between academia, governments, and industry is emphasized. In the case presented by Granquist and Nilsson, Chap. 8, a code of conduct was developed as a tool for management. This, suggest the authors, is one way to mitigate the negative impacts of tourism on local communities and their natural environment. Another way to minimize impacts, propose O'Neill et al. (Chap. 6, this volume), is through environmental impact assessment of human activities; in this particular study, the authors concluded that environmental impacts caused by tourism can be minimized and delimited when properly managed. In case of German tourism activity in Antarctica, Krakau and Herata (Chap. 7, this volume) suggest that nations can also demand that tourism activities, even if they are undertaken outside of the state's border, should be authorized and subject to control mechanisms from the home state. The call for similar action by other states to immediately enact similar and more forceful mechanisms is warranted, especially when one considers the increasing interests in the polar norths. Bastmeijer and Lamers (Chap. 5, this volume) provide an insight in why it is sometimes difficult to overcome the governance challenges outlined by Krakau and Herata. In their chapter, Bastmeijer and Lamers illustrate how differing values and perceptions of human interactions with nature can influence management strategies and in some cases prevent consensus from being reached on certain topics. It is here important to highlight that even if there appears to be small differences in the overarching discourses regarding nature between the nations involved, the de facto situation might be totally divergent.

15.3 Community Implications

Traditional customs and ways of life are also in many ways impacted by increasing tourism needs. The globalization of values and mobilities introduces new areas of potential conflicts between locals and tourists. In Sweden, Finland, and Norway, a custom such as the right of public access, as discussed by Tuulentie and Rantala (Chap. 12), has become the embodiment of this struggle. For example, the authors ask the following: what will happen to "everyman's right" when a "hyper" mobile modern population and new possibilities for non-extractive economic activity cause conflicts with private landowners in resource-rich polar areas?

A community perspective regarding the proposed establishment of national park in Sweden demonstrates the north–south dichotomy associated to the proposed establishment of protected areas in polar areas (Müller, Chap. 13, this volume). As Müller suggests, despite claims of potential economic benefits generated through tourism, provided by proponents (often located in Southern Sweden), it is quite common for locals to question, resist, and, in this particular context, reject this narrative. In terms of the representation of local communities and indigenous people in tourism developments, Lemelin and colleagues (Chap. 14, this volume) focus upon the appeal of dark and warfare tourism. In this particular instance, the authors illustrate how indigenous peoples often are erased in research and in the interpretation of warfare tourism. The omission of indigenous peoples from these areas reinforces colonial narratives and the conquest of the arctic by European explorers, resulting in a cultural dissonance. That said, examples of how cultural dissonance is being addressed in New Zealand do exist and could with some willingness by federal agencies be incorporated in the management of similar sites in the North American Arctic and elsewhere.

15.4 Going Forward

As suggested by Maher (Chap. 2, this volume), research on polar tourism and in tourism more generally has, not to date, encouraged or incorporated theoretical frameworks. Therefore, Maher argues, there needs to be better theoretical linkages and theory building within the field of tourism. The incorporation of theory in tourism is the only way to change and challenge what Viken (Chap. 3) terms "the Anglo-American hegemony of academic writing about Arctic tourism." That in many ways leads to researchers posing the wrong questions and drawing the wrong conclusions.

Tourism in the Arctic has for long been established as one important avenue for regional development by governments, and adding to this, an increasing number of tourists also travel to these areas. Tourism development issues and the impact of tourism in sparsely populated and often peripheral and marginal areas are therefore topics in need for further research, not only as case studies but also regarding the theoretical and methodological contributions that these areas offer. This volume does not separate itself from other books on polar areas in that it does not include a deeper analysis of the economic prospects and pitfalls in a theoretical framework that allows for comparison or theoretical development suited for the polar region and the communities therein. It does however showcase collaborative research approaches conducted with and for local communities. The co-creation and co-dissemination of action research findings is something which is still lacking in polar tourism research. While the roles of local communities (indigenous and nonindigenous) are recognized in this volume, the roles of genders and youth in polar areas are virtually absent and should be addressed in future polar tourism research.

Future polar tourism studies should also contemplate using a system approach to regional economic development, in which institutions of varying kinds, social and cultural elements, as well as economic possibilities are included into a regional (at minimum) arena. An innovation system approach is one way of including polar peoples into the theoretical discussions on economic development from and entrepreneurship in tourism. In research on tourism innovation systems, theories of economic restructuring in natural resource dependent and marginal regions are not commonplace. An innovation system approach to Arctic and Antarctic tourism development brings an extra dimension to the complexity of the analysis of the success and failure of destinations, not least owing to the conditions of sparsely populated peripheral areas with increasing tourism, the threat of climate change and globalization, particularly in the northern but increasingly in the southern hemisphere, too.

In the end, it can be concluded that research on polar tourism still is in its infancy. However, since polar change seems to accelerate and interest in polar tourism grows, further research efforts are pertinent. This should be conducted not only for the people of the far north and south trying to make a living out of polar tourism but also for those tourists who long for polar places and extremes in order to get a positive experience of a lifetime.

Index

D.K. Müller et al. (eds.), *New Issues in Polar Tourism: Communities, Environments, Politics*, DOI 10.1007/978-94-007-5884-1,

MIX
Papier aus verantwortungsvollen Quellen
Paper from responsible sources
FSC® C105338

If you have any concerns about our products,
you can contact us on
ProductSafety@springernature.com

In case Publisher is established outside the EU,
the EU authorized representative is:
Springer Nature Customer Service Center GmbH
Europaplatz 3, 69115 Heidelberg, Germany

Printed by Libri Plureos GmbH
in Hamburg, Germany